SUSTAINABLE CRITICAL INFRASTRUCTURE SYSTEMS

A FRAMEWORK FOR MEETING 21ST CENTURY IMPERATIVES

Report of a Workshop

Toward Sustainable Critical Infrastructure Systems:
Framing the Challenges Workshop Committee

Board on Infrastructure and the Constructed Environment
Division on Engineering and Physical Sciences

NATIONAL RESEARCH COUNCIL
OF THE NATIONAL ACADEMIES

THE NATIONAL ACADEMIES PRESS
Washington, D.C.
www.nap.edu

THE NATIONAL ACADEMIES PRESS   500 Fifth Street, N.W.   Washington, DC 20001

NOTICE: The project that is the subject of this report was approved by the Governing Board of the National Research Council, whose members are drawn from the councils of the National Academy of Sciences, the National Academy of Engineering, and the Institute of Medicine. The members of the committee responsible for the report were chosen for their special competences and with regard for appropriate balance.

This report was supported by Sponsor Award No. W912HQ-06-P-0089 between the National Academy of Sciences and the U.S. Army Corps of Engineers; Sponsor Contract No. CMMI-0751285 between the National Academy of Sciences and the National Science Foundation; Grant No. UPTV-8510 between the National Academy of Sciences and the Construction Industry Institute; and the Presidents' Committee of the National Academies (NRC-8868). Any opinions, findings, conclusions, or recommendations expressed in this publication are those of the author(s) and do not necessarily reflect the views of the organizations or agencies that provided support for the project.

International Standard Book Number-13: 978-0-309-13792-8
International Standard Book Number-10: 0-309-13792-6

Additional copies of this report are available from the National Academies Press, 500 Fifth Street, N.W., Lockbox 285, Washington, DC 20055; (800) 624-6242 or (202) 334-3313 (in the Washington metropolitan area); Internet, http://www.nap.edu.

Printed in the United States of America

THE NATIONAL ACADEMIES
Advisers to the Nation on Science, Engineering, and Medicine

The **National Academy of Sciences** is a private, nonprofit, self-perpetuating society of distinguished scholars engaged in scientific and engineering research, dedicated to the furtherance of science and technology and to their use for the general welfare. Upon the authority of the charter granted to it by the Congress in 1863, the Academy has a mandate that requires it to advise the federal government on scientific and technical matters. Dr. Ralph J. Cicerone is president of the National Academy of Sciences.

The **National Academy of Engineering** was established in 1964, under the charter of the National Academy of Sciences, as a parallel organization of outstanding engineers. It is autonomous in its administration and in the selection of its members, sharing with the National Academy of Sciences the responsibility for advising the federal government. The National Academy of Engineering also sponsors engineering programs aimed at meeting national needs, encourages education and research, and recognizes the superior achievements of engineers. Dr. Charles M. Vest is president of the National Academy of Engineering.

The **Institute of Medicine** was established in 1970 by the National Academy of Sciences to secure the services of eminent members of appropriate professions in the examination of policy matters pertaining to the health of the public. The Institute acts under the responsibility given to the National Academy of Sciences by its congressional charter to be an adviser to the federal government and, upon its own initiative, to identify issues of medical care, research, and education. Dr. Harvey V. Fineberg is president of the Institute of Medicine.

The **National Research Council** was organized by the National Academy of Sciences in 1916 to associate the broad community of science and technology with the Academy's purposes of furthering knowledge and advising the federal government. Functioning in accordance with general policies determined by the Academy, the Council has become the principal operating agency of both the National Academy of Sciences and the National Academy of Engineering in providing services to the government, the public, and the scientific and engineering communities. The Council is administered jointly by both Academies and the Institute of Medicine. Dr. Ralph J. Cicerone and Dr. Charles M. Vest are chair and vice chair, respectively, of the National Research Council.

www.national-academies.org

TOWARD SUSTAINABLE CRITICAL INFRASTRUCTURE SYSTEMS: FRAMING THE CHALLENGES WORKSHOP COMMITTEE

DAVID J. NASH, *Chair*, Dave Nash and Associates, Birmingham, Alabama

E. SARAH SLAUGHTER, *Vice Chair*, Massachusetts Institute of Technology, Cambridge

MASSOUD AMIN, University of Minnesota, Minneapolis

MARK EBERHART, Colorado School of Mines, Golden

HENRY J. HATCH, U.S. Army Corps of Engineers (retired), Oakton, Virginia

SUE McNEIL, University of Delaware, Newark

ROBERT PRIETO, Fluor Corporation, Princeton, New Jersey

GARRET P. WESTERHOFF, Malcolm Pirnie, Inc. (chairman emeritus), Fairlawn, New Jersey

STAFF

LYNDA STANLEY, Study Director

KEVIN LEWIS, Senior Program Officer

HEATHER LOZOWSKI, Financial Associate

TERI THOROWGOOD, Administrative Coordinator

Preface

Much has been written in the past 20 years about the deteriorating condition and quality of the nation's roads and bridges and its power and water systems and about the trillions of dollars that it will take to fix them. The issues, however, are much more complex: How we as a nation choose to renew our infrastructure systems in the coming years will help determine the quality of life for future generations. It will also help determine our success in meeting other national challenges, including those of remaining economically competitive, reducing our dependence on imported oil, and dealing with issues related to global climate change, national security, and disaster resilience.

Many of the ideas and much of the information in this report were generated at a workshop held May 7 and 8, 2008, at the National Academies in Washington, D.C. In the time between the workshop and the completion of this report, global and domestic financial markets and systems have faltered in ways reminiscent of the 1930s. An economic stimulus package has been approved that includes funding for some infrastructure projects. Nonetheless, the message of this report—that a paradigm shift is needed in how the nation thinks about, builds, operates, and invests in critical infrastructure systems in the long term—is more relevant than ever.

The terms *critical infrastructure* and *sustainability* are frequently used in different contexts. In this report, *critical infrastructure* refers to the water, wastewater, power, transportation, and telecommunications systems—sometimes called lifeline systems—without which buildings, emergency response systems, dams, and other infrastructure cannot operate as intended. *Sustainability* is broadly defined to mean the ability of systems to meet the needs of current and future generations by being physically resilient, cost-effective, environmentally viable, and socially equitable.

In recent years, many organizations and individuals have called attention to the deteriorating condition of our infrastructure and have offered solutions for renewing it. At this time, however, the United States does not have a vision or concept of, or stated objectives for, the future configuration, level of performance, or level of services that critical infrastructure systems should provide. Current local, regional, and national policies, processes, and practices are structured to treat these systems as stand-alone entities even though they are interdependent and the solutions chosen to "fix" one system will affect the others. Lacking an overall strategy for infrastructure renewal and focusing on one system, one issue, or one problem at a time, the nation runs the risk of wasting increasingly scarce resources and of creating new problems for future generations.

In addition, the nation has not yet fully considered how critical infrastructure system renewal can be leveraged to help meet other national challenges, especially economic competitiveness, global climate change, national security, energy independence, and disaster resiliency. Although not intuitive, the linkages between these challenges and critical infrastructure systems are real and significant. Power, transportation, water, and telecommunications are the basis for producing and delivering goods and services that are key to economic competitiveness, emergency response and recovery, and quality of life. These systems also account for 69 percent of the nation's total energy use and for more than 50 percent of the greenhouse gas emissions linked to global climate change (EIA, 2008b).

Because the infrastructure systems in the United States are deteriorating and require significant reinvestment, now is the time to conduct a fundamental reexamination of the value and purposes of critical infrastructure systems, their interdependencies, and the policies, processes, and procedures that guide their planning, construction, operation, and investment. A world of new possibilities and approaches to infrastructure renewal will open up if we choose to think about critical infrastructure more holistically, in terms of the services that these systems provide—water, wastewater removal, power, mobility, and connectivity—and as part of a strategy for meeting other national imperatives. To paraphrase Albert Einstein, the significant problems we face cannot be solved at the same level of thinking we were at when we created them.

The difficulties inherent in achieving a paradigm shift in the ways that Americans think about and invest in infrastructure cannot be solved by any one group or single strategy, nor do they need to be. New approaches, strategies, and ideas for providing and financing infrastructure-related services are being generated at the grassroots level; in universities; by local, state, and federal governments; by regional and not-for-profit organizations; and in the private sector. New technologies and materials can allow people to provide power and mobility using alternatives to imported oil, to save water and other dwindling resources, to reduce greenhouse gas emissions, and to create infrastructure systems that are more durable, reliable, resilient, and cost-effective.

The challenge is to marshal the knowledge, creativity, financial resources, and energy of a diverse array of individuals, interests, and organizations to develop new concepts, approaches, and strategies for critical infrastructure renewal. A framework is needed to allow these groups to work collaboratively in a structured way in order to develop practical, cost-effective solutions based on objective, evidence-based information.

This report provides the elements of such a framework. It calls for a broad and compelling vision for the future; a focus on providing essential services; recognition of the interdependencies of critical infrastructure systems to enable the achievement of multiple objectives; collaborative, systems-based approaches to leverage available resources; and performance measures to provide transparency about infrastructure investments.

Bringing the various stakeholder groups together to undertake this challenging endeavor first requires a call to action from leaders at the highest levels of government, academia, and industry. Succeeding in this effort will require imagination; a commitment to collaborate in the search for innovative, cost-effective, long-term solutions; and the will to stay the course.

David J. Nash, *Chair*
Toward Sustainable Critical
Infrastructure Systems:
Framing the Challenges
Workshop Committee

The Board on Infrastructure and the Constructed Environment (BICE) was established by the National Research Council (NRC) in 1946 as the Building Research Advisory Board. BICE brings together experts from a wide range of scientific, engineering, and social science disciplines to discuss potential studies of interest, develop and frame study tasks, ensure proper project planning, suggest possible reviewers for reports produced by fully independent ad hoc study committees, and convene meetings to examine strategic issues. Only those board members who were appointed to the Toward Sustainable Critical Infrastructure Systems: Framing the Challenges Workshop Committee were asked to endorse the committee's conclusions or recommendations.

Previous BICE reports on the nation's critical infrastructure systems include the following: *In Our Own Backyard: Principles for Effective Improvement of the Nation's Infrastructure* (1993); *Toward Infrastructure Improvement: An Agenda for Research* (1994); *Measuring and Improving Infrastructure Performance* (1995); *The Challenge of Providing Future Infrastructure in an Environment of Limited Resources, New Technologies, and Changing Social Paradigms: Proceedings of a Colloquium* (1995); and *Financing Tomorrow's Infrastructure: Challenges and Issues: Proceedings of a Colloquium* (1996). These reports and additional information about BICE can be obtained online at http://www.nationalacademies.org/bice.

Acknowledgments

The authoring committee acknowledges the significant contributions made by all of the workshop participants, who willingly and enthusiastically volunteered their time and ideas. Special recognition is owed to Hans Van Winkle and Wayne Crew of the Construction Industry Institute and to Jesus de la Garza, formerly with the National Science Foundation, whose conversations in 2006 were the genesis of the workshop.

This report has been reviewed in draft form by individuals chosen for their diverse perspectives and technical expertise, in accordance with procedures approved by the National Research Council's Report Review Committee. The purpose of this independent review is to provide candid and critical comments that will assist the institution in making its published report as sound as possible and to ensure that the report meets institutional standards for objectivity, evidence, and responsiveness to the study charge. The review comments and draft manuscript remain confidential to protect the integrity of the deliberative process. We wish to thank the following individuals for their review of this report:

George Bugliarello, Polytechnic Institute of New York
    University,
John L. Casti, Complexica,
Nancy Rutledge Connery, Independent Consultant,
    Woolwich, Maine,

Pablo Garcia, Sandia National Laboratories,
Paul H. Gilbert, Parsons Brinckerhoff Quade & Douglas,
James Elliott Moore II, University of Southern California,
Yossi Sheffi, Massachusetts Institute of Technology,
Kumares C. Sinha, Purdue University, and
Robert D. Yaro, Regional Plan Association.

Although the reviewers listed above have provided many constructive comments and suggestions, they were not asked to endorse the conclusions or recommendations, nor did they see the final draft of the report before its release. The review of this report was overseen by Richard Wright, National Institute of Standards and Technology (retired). Appointed by the National Research Council, he was responsible for making certain that an independent examination of this report was carried out in accordance with institutional procedures and that all review comments were carefully considered. Responsibility for the final content of this report rests entirely with the authoring committee and the institution.

Contents

Summary

For the people of the United States, the 20th century was one of unprecedented population growth, economic development, and improved quality of life. As the population grew by 205 million (U.S. DOC, 2008), Americans invested huge sums of money to meet a range of social, economic, and political imperatives or urgent needs. These imperatives included providing for public health, safety, and comfort, national security, economic expansion, and job growth. Meeting these imperatives required the construction of water, wastewater, and power systems, as well as roads, railways, airports, and telecommunications systems. As these imperatives were met, cities and regions were developed, great swathes of forest and land were cleared, rivers were controlled and channeled, and renewable and nonrenewable resources were harvested, extracted, and productively used.

The critical infrastructure systems—water, wastewater, power, transportation, and telecommunications[1]—built in the 20th century have become so much a part of modern life that they

---

[1]Infrastructure systems have been defined differently by different groups. For example, the Department of Homeland Security's National Infrastructure Protection Plan identifies 18 types of infrastructure (DHS, 2009). The American Society of Civil Engineers' "Report Card for America's Infrastructure" identifies 15 types (ASCE, 2005). In this report, *critical infrastructure systems* are defined as power, water, wastewater, telecommunications, and transportation systems. These five systems are the lifelines without which other types of infrastructure (e.g., banking and finance, government facilities, schools) cannot operate as intended.

are taken for granted. Today, Americans expect reliable power, clean drinking water, uncomplicated travel, and easy communications. Businesses and industries rely on infrastructure systems to transport raw materials, manufacture products, deliver goods, share ideas, and conduct transactions. The demands on these critical systems will continue to grow: By 2030, 60 million more Americans, in addition to new businesses and employers, will expect these systems to deliver essential services (U.S. DOC, 2008).

The quality of life for Americans in the 21st century will depend in part on whether the nation's critical infrastructure systems can meet such expectations. At the same time, other national challenges that will affect quality of life include, but are not limited to, the following:

- Remaining economically competitive with the European Union, China, India, and other economic powers;
- Reducing U.S. dependence on imported oil;
- Reducing the greenhouse gas emissions linked to global climate change;
- Protecting the environment and conserving increasingly scarce natural resources, including potable water; and
- Developing the capacity to withstand and recover quickly from natural and human-made disasters.[2]

Meeting these challenges requires critical infrastructure systems that are efficient, reliable, and cost-effective. If the United States is to remain a global economic leader, new methods and corridors will be needed to move goods and services efficiently to and from domestic and international markets, including Canada and Mexico. Reducing petroleum imports will require strategies to reduce demand and may require new infrastructure—microgeneration facilities, power plants, and distribution networks—for the efficient generation, storage, distribution, and use of power from alternative sources of energy.

Today, electric power and transportation account for 40 percent and 29 percent, respectively, of the nation's total annual energy use, and together they account for more than 50 percent of the greenhouse gas emissions linked to global climate change (EIA, 2008b). Significantly reducing greenhouse gas emissions

---

[2]Imperatives besides those listed include health care and the solvency of the Social Security system.

will require that power and mobility be provided in new ways using new systems and technologies. Limiting deaths and injuries, property losses, impacts on ecosystems, and recovery time after natural or human-made disasters requires robust and resilient infrastructure systems. The processes and materials used to renew or create critical infrastructure systems will be significant factors in achieving or failing to achieve environmental sustainability.

Large segments and components of the nation's critical infrastructure systems are now 50 to 100 years old. Their performance and condition are deteriorating, as evidenced by transportation congestion, air and water pollution, and increasing instances of power and other service disruptions (ASCE, 2009; Amin, 2008). In recent years, both public and private organizations have called for improvements in critical infrastructure systems, often recommending massive financial investments.[3] However, approaching infrastructure renewal by continuing to use the same processes, practices, technologies, and materials that were developed in the 20th century will likely yield the same results: increasing instances of service disruptions, higher operating and repair costs, and the possibility of catastrophic, cascading failures such as those in New Orleans in 2005 following Hurricane Katrina. If the nation is to meet some of the important challenges of the 21st century, a new paradigm for the renewal of critical infrastructure systems is needed.

BACKGROUND

In 2007, the National Research Council (NRC) appointed an ad hoc committee of experts (Appendix A provides biosketches of the committee members) to identify and frame fundamental challenges in moving toward critical infrastructure systems that are physically, socially, economically, and environmentally sus-

---

[3]See, for example, "America's Infrastructure: Ramping Up or Crashing Down" (Katz et al., 2007); "Guiding Principles for Strengthening America's Infrastructure" (CSIS, 2006); "A New Bank to Save Our Infrastructure" (Ehrlich and Rohatyn, 2008); "Time for an Infrastructure Overhaul" (Little, 2007); "It's Time to Rebuild America" (Rohatyn and Rudman, 2005); "Report Card for America's Infrastructure—2009" (ASCE, 2009); and "Main Street, Not Wall Street, Should Fix Crumbling Infrastructure" (Sebelius and Stern, 2008).

tainable.[4] As its principal data-gathering activity, the committee conducted a workshop on May 7 and 8, 2008, in Washington, D.C., bringing together approximately 50 experts from government, academia, and the private sector (Appendix B presents the list of participants, and Appendix C includes the workshop agenda). This report summarizes the committee's findings based on the workshop outcomes (Appendix D provides a succinct presentation of the outcomes), published materials, and the expertise and experience of its members. It provides a new context for thinking about the purposes and value of critical infrastructure systems: It does so by focusing on the links between some of the imperatives of the 21st century (economic competitiveness, global climate change, reducing U.S. dependence on imported oil, disaster resiliency, and environmental sustainability) and the performance of critical infrastructure systems. The report does not make specific recommendations, but instead it identifies a framework for developing a new paradigm for investing in and renewing critical infrastructure systems in ways that will also help meet other 21st century challenges.

FINDINGS

Renewing and restructuring the nation's critical infrastructure systems to meet some of the important challenges of the 21st century constitute a task radically different from that of building new systems across undeveloped territory. Renewal efforts must take into account an extensive network of existing systems, urban development, ownership patterns, construction processes, management practices, financing mechanisms, and regulatory mandates.

Every year public- and private-sector organizations spend hundreds of billions of dollars to operate and maintain power, water, wastewater, transportation, and telecommunications systems. At least $285 billion was invested in these efforts in 2004 alone (CBO, 2008). Nonetheless, this level of investment has not been adequate, as evidenced by the deteriorating condition of these systems. The resources available to renew and restructure

---

[4]The committee defined *sustainable* as meeting today's economic, social, and environmental needs while enhancing the ability of future generations to meet their economic, social, and environmental needs.

these systems will be limited for the foreseeable future, and investment choices will need to be made.

At a time when many have called for infrastructure renewal in some form and have suggested billions or trillions in investment, there is an important opportunity to fundamentally reexamine the purposes and value of critical infrastructure systems and of the decision-making processes used for investing in them. While daunting, this reexamination can yield a new paradigm from which to develop practical, cost-effective solutions to complex challenges and help meet the needs of future generations.

Some of the ingredients needed to create a new paradigm are available today. Research has yielded technologies for monitoring infrastructure condition and performance, new materials for constructing and repairing infrastructure components, and new knowledge about the interrelated nature of water and wastewater, power, transportation, and telecommunications systems. Self-diagnosing, self-healing, and self-repairing systems can be designed to provide for greater resiliency, fewer long-term service disruptions, and lower life-cycle costs (Amin and Stringer, 2008). An array of financing mechanisms, strategies, plans, and approaches to infrastructure renewal that offer new ways to provide for essential services has been developed through local, state, and regional initiatives.

To date, however, infrastructure-related technological advances, plans, approaches, and community-based initiatives have been ad hoc in nature, often focusing on one issue, one type of system, or one set of solutions. Lacking a national vision or strategy for critical infrastructure renewal and concentrating on single projects, technologies, financing mechanisms, or narrowly defined objectives, ad hoc efforts run the risk of underutilizing or wasting scarce resources and increasing the probability of serious, unintended consequences. A framework is needed to structure these efforts so that ongoing activities, knowledge, and technologies can be aligned and leveraged to help meet multiple national objectives. The essential components of the needed framework are as follows:

- *A broad and compelling vision* that will inspire individuals and organizations to pull together to help meet 21st century imperatives by renewing the nation's critical infrastructure systems. Such a vision would focus on a future of economic competitiveness, energy independence, environmental sustainability, and quality of life, not a legacy of concrete, steel, and cables.

- *A focus on providing the essential services involving water and wastewater, power, mobility, and connectivity*—in contrast to upgrading individual physical facilities—to foster innovative thinking and solutions.
- *Recognition of the interdependencies among critical infrastructure systems* to enable the achievement of multiple objectives and to avoid narrowly focused solutions that may well have serious, unintended consequences.
- *Collaborative, systems-based approaches* to leverage available resources and provide for cost-effective solutions across institutional and jurisdictional boundaries.
- *Performance measures* to provide for greater transparency in decision making by quantifying the links among infrastructure investments, the availability of essential services, and other national imperatives.

An important first step in creating a new paradigm is to bring together those who have an essential stake in meeting 21st century imperatives and who are already involved in sustainable infrastructure efforts. They include infrastructure owners, designers, engineers, financiers, regulators, and policy makers, as well as ecologists, community activists, scientists, and researchers. Working within the framework, experts in such areas could begin to identify a full range of new approaches, technologies, and materials for providing services involving mobility, connectivity, water, wastewater, and power to meet multiple objectives. They could also identify new approaches to the decision making, finance, and operations processes related to critical infrastructure systems. The results of such a gathering could serve to initiate a longer-term, collaborative effort to develop a vision that would provide guidance for developing concepts and objectives for the nation's critical infrastructure systems and then to identify the policies, practices, and resources required to implement them. The results could be critical infrastructure systems that are physically resilient, cost-effective, socially equitable, and environmentally sustainable for the next 50 years.

1 Why Critical Infrastructure Systems Matter

LIFELINE SYSTEMS

The term *infrastructure* has been used many different ways to include a variety of components. In this report, *critical infrastructure systems* are defined as the water, wastewater, power, transportation, and telecommunications systems without which buildings, emergency response systems, and other infrastructure cannot operate as intended. They are the "lifeline systems" that physically tie together metropolitan areas, communities, and neighborhoods, and facilitate the growth of local, regional, and national economies. These interdependent systems work together to provide the essential services of a modern society:

- Water for a vast array of needs, including drinking, washing, cooking, firefighting, farming, and sanitation, as well as for manufacturing, industrial, and mining processes;
- Power for numerous uses, including heat, light, refrigeration, cooking, food processing, and security purposes; the production of durable goods; and the operation of oil and gas refineries, the Internet, television, and appliances;
- Mobility for people, materials, goods, and services to and from workplaces, markets, schools, recreational facilities, and other destinations;
- Connectivity for purposes of communication, public safety, emergency services, financial transactions, and

for the control and monitoring of other infrastructure components.

Opinions among economists vary about the role of public spending for infrastructure as a means of creating jobs and equalizing opportunity. However, economists generally agree that (1) infrastructure and its quality affect behavior with respect to location—that is, where people, activities, and businesses are located or willing to locate—which in turn affects economic growth, land use, and quality of life; and (2) it is difficult to achieve high rates of productivity in the absence of quality infrastructure (Gramlich, 1994). Thus, the efficiency, reliability, and resiliency of critical infrastructure systems affect many aspects of society, including the following:

- The costs of food, durable goods, and consumer goods;
- The competitiveness of U.S. services and goods in the global market;
- The health, safety, and well-being of citizens;
- The quality of life in communities;
- The availability and reliability of power and the maintenance of life-support systems;
- The travel time required for people to go from home to work or other destinations and for the efficient transport of goods and services;
- The reliability and speed of telecommunications;
- The speed and effectiveness of communications about actions to be taken during natural and human-made disasters (e.g., regarding evacuation and safe harbors);
- The time, cost, and extent of recovery for communities following such disasters.

Critical infrastructure systems also affect the quality of the environment and the availability of natural resources for other uses. Electric power and transportation account for 40 percent and 29 percent, respectively, of the nation's total annual energy use; together they account for more than 50 percent of the greenhouse gas emissions linked to global climate change (EIA, 2008b).

Critical infrastructure systems are built to provide services to several generations over several decades. These systems have become so integrated into modern life that they are taken for granted: Today, Americans expect to have power at the flip of a switch, clean drinking water by turning on a tap, the mobility to travel freely at any time, and the connectivity to communicate instantaneously. Today, in U.S. businesses and industries, it is expected and relied on that the required infrastructure is available to transport raw materials, to manufacture products, to deliver food and durable goods to markets and ports, and to enable the sharing of ideas and the conduct of transactions electronically. By 2030, an additional 60 million Americans and unknown numbers of businesses will have similar demands and expectations for the services provided by these systems (U.S. DOC, 2008).

EFFECTS OF DETERIORATING CONDITIONS

Although the nation invested heavily in the design, construction, and operation of these systems, it has not invested the funds necessary to keep these systems in good condition or to upgrade them to meet the demands created by a growing and shifting population. Large segments and components of the nation's water, wastewater, power, transportation, and telecommunications systems are now 50 to 100 years old. Some systems and components are physically deteriorating owing to wear and tear and lack of timely maintenance and repair, which can lead to increasing rates of intermittent and periodic loss of service. For instance, in the United States between 1991 and 2000, 99 separate power outages occurred, affecting at least 50,000 consumers each time. However, between 2001 and 2005, there were 150 outages affecting 50,000 or more consumers—that is, there were 50 percent more outages in half the time (Amin, 2008).

The performance of systems is also deteriorating where system capacity is not adequate for the level of use. Each year, for example, every driver spends an average of 25 hours in traffic delays at a cost of $742 in time and fuel (TTI, 2005).

When critical infrastructure systems fail completely, the results can be devastating, as evidenced by the following events:

- *The Northeast power blackout of 2003*, during which 50 million people lost power for up to 2 days, at an estimated cost of $6 billion (Minkel, 2008);

- *Twelve steam pipe explosions in New York City between 1989 and 2007*, which killed several people, disrupted power and commerce, and required costly repairs (Belson and DePalma, 2007);
- *The collapse of the I-35W bridge in Minneapolis, Minnesota, in 2007*, resulting in 13 deaths, numerous injuries, the disruption of commerce for more than 1 year, and the need for a new bridge at a cost of $233 million (Figure 1.1) (MnDOT, 2007); and
- *The levee failures in New Orleans in 2005,* resulting in approximately 1,500 deaths; between $20 billion and $22 billion in property losses; $4 billion to $8 billion in economic losses; $16 billion to $20 billion in emergency assistance (Kates et al., 2006); and economic, social, and environmental effects that are being felt more than 3 years later.

Infrastructure can also fail if subjected to terrorist attack, as on September 11, 2001, with the collapse of the Twin Towers of the World Trade Center in New York City. The National Infrastructure Protection Plan developed by the Department of Homeland Security states:

> Protecting and ensuring the resiliency of the critical infrastructure and key resources (CIKR) of the United States is essential to the Nation's security, public health and safety, economic vitality, and way of life. Attacks on CIKR could significantly disrupt the functioning of government and business alike and produce cascading effects far beyond the targeted sector and physical location of the incident. Direct terrorist attacks and natural, manmade, or technological hazards could produce catastrophic losses in terms of human casualties, property destruction, and economic effects, as well as profound damage to public morale and confidence. Attacks using components of the Nation's CIKR as weapons of mass destruction could have even more devastating physical and psychological consequences (DHS, 2009, p. 1).

In summary, critical infrastructure systems matter because they directly affect—both positively and negatively—the daily lives of all Americans. These systems provide the essential services for health, comfort, and prosperity. However, their deteriorating levels of condition and performance routinely

FIGURE 1.1 The scene of the collapse of the I-35W bridge in Minneapolis, Minnesota, in 2007. SOURCE: Minnesota Department of Transportation. Available at http://www.dot.state.mn.us/i35wbridge/photos/aerial/aug-16/index.htm.

inconvenience individuals, pose risks to communities during and after emergencies, and inhibit the nation's capacity to move goods and services efficiently to domestic and international markets. How the nation chooses to renew these systems will have a direct bearing on local, regional, and national economies and on the quality of life for more than 300 million Americans. Critical infrastructure system renewal will also have a direct impact on how the nation meets some other imperatives of the 21st century, as described in Chapter 2.

ORIGIN AND BACKGROUND OF THE REPORT

This report grew out of discussions held in 2006 and 2007 among current and former staff of the National Science Foundation, the Construction Industry Institute, the National Institute of Standards and Technology, and the Board on Infrastructure and the Constructed Environment of the National Research

Council (NRC). In 2007, the NRC appointed an ad hoc committee of experts (Appendix A provides biosketches of the committee members) to identify and frame fundamental challenges in moving toward critical infrastructure systems that are physically, socially, economically, and environmentally sustainable.[1]

As its principal data-gathering activity, the committee conducted a workshop on May 7 and 8, 2008, in Washington, D.C., bringing together approximately 50 experts from government, academia, and the private sector (Appendix B presents the list of participants). The committee developed a draft set of critical infrastructure-related challenges to serve as the starting point for a series of breakout sessions during the workshop. The participants commented on and modified the draft challenges and identified potential lines of inquiry—policies, processes, financing mechanisms, technologies, materials, and research—that might be used to address the challenges (Appendix C contains the workshop agenda and a list of the draft challenges).

This report summarizes the committee's findings based on the workshop outcomes (Appendix D provides a succinct presentation of the outcomes), published materials, and the expertise and experience of its members. It provides a new context for thinking about the purposes and value of critical infrastructure systems: It does so by focusing on the links between some of the imperatives of the 21st century (economic competitiveness, global climate change, reducing U.S. dependence on imported oil, disaster resiliency, and environmental sustainability) and the performance of critical infrastructure systems. The report focuses on broad concepts; others have written about these issues in much greater detail in various studies and articles. The report does not make specific recommendations, but instead it identifies a framework for developing a new paradigm for investing in and renewing critical infrastructure systems in ways that will also help meet other 21st century challenges.

---

[1]The committee defined *sustainable* as meeting today's economic, social, and environmental needs while enhancing the ability of future generations to meet their economic, social, and environmental needs.

2 Meeting 21st Century Imperatives with 20th Century Infrastructure Systems

The 20th century was one of unprecedented economic growth and improved quality of life for Americans. As the nation's population more than tripled, from 76 million in 1900 to 281 million in 2000 (U.S. DOC, 2008), huge investments were made to build the critical infrastructure systems required to meet a range of social, economic, and political imperatives. Water and wastewater systems were built to support population growth, industrial growth, and public health. Power systems were built to heat and light homes, schools, and businesses and to energize communications and factories. Roads, railroads, and airports were built to support mobility and commerce. And telecommunications systems were built to provide connectivity within neighborhoods and across the world.

In the 21st century, critical infrastructure systems will play an essential role in meeting other urgent national needs or imperatives, including the following:

- Remaining economically competitive with the European Union, China, India, and other economic powers;
- Reducing U.S. dependence on imported oil;
- Reducing the greenhouse gas emissions linked to global climate change;
- Protecting the environment and conserving increasingly scarce natural resources, including potable water; and

- Developing the capacity to withstand and recover quickly from natural and human-made disasters.

The links between critical infrastructure systems and these 21st century imperatives are not always obvious. However, they are real and significant.

ECONOMIC COMPETITIVENESS

Throughout much of the 20th century, the United States was the global economic leader, and it remains so today. However, new technologies, political changes, and other factors have led to greater economic competition among nations, new production centers, and new trading patterns, all of which have implications for U.S. competitiveness in the future. The Internet and other technologies have changed the structure of businesses and the location of production centers around the world (Mongelluzzo, 2008). The development of "megaships" for transporting containerized goods, implementation of the North American Free Trade Agreement (NAFTA), and other major factors are changing trading patterns among nations. The fall of communism in the Soviet Union and Eastern Europe and the emergence of the European Union, China, and India as economic powers have resulted in greater wealth and consumer demand throughout the world (Gallis, 2008). For the United States, international trade (imports and exports) increased yearly between 1997 and 2005 as a proportion of the gross domestic product, a trend that is projected to continue through 2030 (Figure 2.1).

A key enabler of global trade is the "increasingly complex just-in-time supply chain logistics system, which depends, in turn, on reliable power, mobility, and water" (Doshi et al., 2007, p. 4). Critical infrastructure systems, in fact, provide the foundation for producing and moving goods and services to seaports, airports, and shipping terminals for export to other countries.

The primarily east-west configuration of the nation's highways, railways, and shipping terminals reflects the trading patterns of the 20th century. Food, vehicles, and other goods were primarily produced in the center of the country and transported to major cities on the East, West, and Gulf Coasts for domestic consumption and for shipment to Europe and Asia.

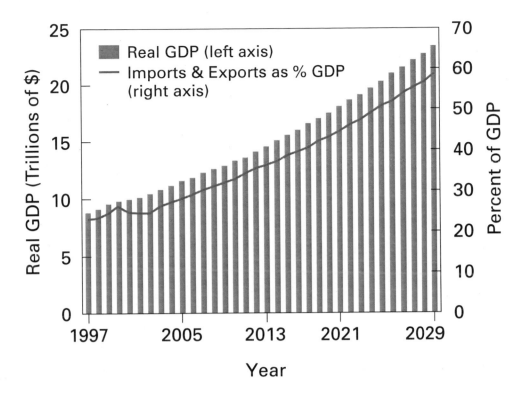

FIGURE 2.1 Sum of imports and exports compared with U.S. gross domestic product (GDP) as projected through 2030. SOURCE: TRB (2006).

As new economic powers emerge, global trading patterns are changing. New ports are developing along the west coast of Mexico from which goods are shipped north to Los Angeles, San Francisco, and Seattle by ground and to Chicago, Detroit, and Toronto by air (Gallis, 2008). On the East Coast, goods are being transported from Halifax in Canada south to New York and the Gulf Coast. Canada and Mexico also supply a significant portion of the petroleum used in the United States. Trade routes from Southeast Asia across the Indian Ocean, into the Red Sea, and across the Mediterranean Sea mean that Asian goods can be directly delivered in containers to East Coast cities in the United States instead of being shipped to the West Coast and transported across the country (Gallis, 2008). The expansion of the Panama Canal by 2014 to accommodate megaships will allow Asian goods more direct access to East Coast ports (Mongelluzzo, 2008) (Figure 2.2).

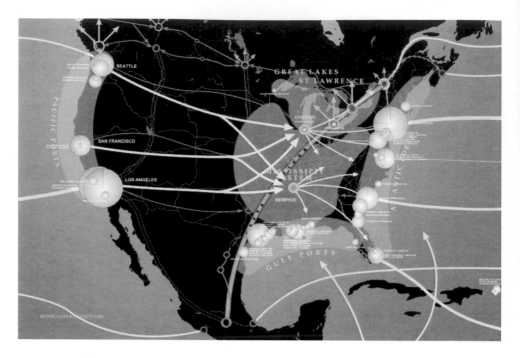

FIGURE 2.2 The changing global economy is changing shipping patterns. SOURCE: Gallis (2008).

The primarily east-west configuration of U.S. critical infrastructure systems does not reflect the north-south trade patterns with Canada and Mexico. Increased trade following the adoption of NAFTA, combined with new security requirements, "has caused significant congestion and cost increases at border crossings with Mexico and Canada and on corridors serving NAFTA markets" (TRB, 2006, pp. 2-3). A separate but related issue is that "West Coast ports may be unable to handle the staggering projected growth in Asian trade over the next 20 years—even with significant increases in port productivity—because of landside constraints on rail and highway systems" (TRB, 2006, p. 2).

To improve their competitiveness, other economic powers have developed integrated strategies for economic growth that include infrastructure as a key component. In 1986, the Ministry of Science and Technology of the People's Republic of China launched a national high-technology research and development plan "to meet the global challenges of new technology revolution and competition" (MSTPRC, 2006). The program is now in its 10th Five-Year Plan period. The European Union Treaty "obliges the Community to contribute to the organization and development of Trans-European Networks (TENs) in the areas of transport, telecommunications and energy supply infrastructure . . . to serve the objectives of a smooth functioning Single Market . . ." (EC, 1999, p. 14). The United States, in contrast, does not have a strategy to link its infrastructure to its global competitiveness.

Domestically, congested highways, airports, and shipping terminals also impede the efficient movement of raw materials, meat, produce, and durable goods destined for local and regional markets. It has been estimated that highway congestion costs Americans approximately $65 billion per year (2005 dollars) and wastes 2.3 billion gallons of gasoline (TRB, 2006). The additional costs incurred by such congestion increase the costs of food, fuel, and other commodities for every consumer. If the United States is to remain as economically competitive as possible, more efficient methods to transport goods and services and additional corridors may be needed. New corridors or infrastructure components in turn could have significant environmental and land use impacts unless they are fully evaluated and carefully planned.

REDUCING U.S. DEPENDENCE ON IMPORTED OIL

While 42 percent of the petroleum used in the United States comes from domestic sources, 58 percent is imported (EIA, 2008a). The majority of imported oil comes from Canada (18 percent), the Persian Gulf countries (16 percent), Mexico (11 percent), Venezuela (10 percent), and Nigeria (8 percent) (EIA, 2008a). Some of these countries are politically unstable, and transporting supplies to market involves vulnerable points that are subject to disruption (NRC, 2008a). With demand for energy increasing around the world in combination with limited supplies of oil, prices for petroleum are likely to rise over the long term. Decreasing the nation's dependence on imported oil has implications for national security as well as for consumers' pocketbooks.

Reducing the level of imported petroleum will depend in part on strategies to reduce overall demand (for example, by means of more fuel-efficient cars and greater reliance on public transportation); on whether the United States is able to efficiently generate, store, distribute, and use power from domestically available, alternative sources of energy; and on other measures.

Opportunities exist to produce power from wind, the Sun, hydrogen, and other sources of energy. The construction of new infrastructure—microgeneration facilities, power plants, and distribution networks—may be required. Some alternative energy power projects have been developed—such as those converting the methane gas produced by landfills to energy—and many have been proposed. However, they are being implemented on a case-by-case basis in the absence of an overarching strategy.

A range of demand-side and supply-side strategies are available that could lead to a reduction in the national demand for imported oil. Each brings with it a host of implications for future development and future generations. Any pursuit of narrowly focused objectives and one-dimensional strategies, however, could lead to serious, unintended consequences. For example, the focus on producing ethanol derived from corn kernels as a biofuel to reduce the demand for imported oil has had unforeseen impacts on the cost of corn for food products and has not fully taken into account the impacts on water availability, water quality (NRC, 2008c), and other factors. Ad hoc development of new infrastructure systems could lead to redundancies in some areas, a lack of service in others, the waste of valuable resources, and adverse environmental impacts.

To the extent that new systems or components of systems are developed, they will require substantial public- and private-sector investments. Typically, major infrastructure projects take 10 to 20 years or more to plan, approve, obtain needed permits, fund, and build. Even with the careful planning, design, and siting that promise to mitigate environmental impacts, local opposition is likely to arise, a phenomenon sometimes referred to as NIMBYism (for "not in my backyard"). Coordinated action across political jurisdictions and stakeholder groups as well as broad public support will be needed to develop cost-effective infrastructure systems required to deliver energy from alternative sources in the next 10 to 20 years. Coordinated action will be difficult to achieve in the absence of an overarching concept or objectives for critical infrastructure systems.

GLOBAL CLIMATE CHANGE

Scientists predict that global climate change—higher temperatures and extremes of precipitation—will result in more extreme instances of drought and flooding, as well as tropical storms of increased intensity and rising sea levels (NRC, 2008b). These changes will profoundly affect agriculture and forest productivity, ecosystems, water, and other resources, which will in turn affect societies and communities. If current weather trends continue as predicted, rising sea levels and greater storm surges will have significant consequences for shipping ports, terminals, and the infrastructure systems of some of the country's largest cities[1] and other coastal communities (NRC, 2008b). In large portions of the country, more-intense, longer-lasting droughts will affect the availability of water for drinking, irrigation, fire suppression, and sanitation. Record levels of precipitation, in contrast, will result in more instances of flooding, land erosion, and the undermining of roads and other infrastructure systems.

Greenhouse gas emissions (carbon dioxide, methane, nitrous oxide, and other gases) are a factor in higher temperatures. These emissions are produced by the burning of fossil fuels, including oil, natural gas, and coal; by wastewater treatment plants; by the production of cement and other materials; and by other human activities. Electric power and transportation alone accounted for more than 50 percent of the nation's total greenhouse gas emissions in 2007 (EIA, 2008b).

Water resources and systems will also be affected by climate change. According to the Intergovernmental Panel on Climate Change (IPCC):[2]

> Climate change affects the function and operation of existing water infrastructure—including hydropower, structural flood defences [sic], drainage and irrigation systems—as well as water management practices. . . . Current water management practices may not be robust enough to cope with the impacts of climate change on water supply reliability, flood risk, health, agriculture, energy, and aquatic ecosystems. . . . Adaptation options designed to ensure water supply during average and drought conditions require integrated demand-side as well as supply-side strategies (Bates et al., 2008, p. 4).

---

[1]On the East Coast: New York City, Jacksonville, Florida, and Baltimore, Maryland; on the West Coast: Los Angeles, San Diego, and San Francisco, and Seattle, Washington; and on the Gulf Coast: Houston, Texas.

[2]The IPCC is a scientific intergovernmental body set up by the World Meteorological Organization and by the United Nations Environment Programme. Additional information about this organization is available at http://www.ipcc.ch/. Accessed February 12, 2009.

The western United States is one of the areas of the world that is "particularly exposed to the impacts of climate change" and is "projected to suffer a decrease of water resources" (Bates et al., 2008, p. 3). Demand-side strategies for mitigating these impacts could include both greater efficiency of water use through recycling and greater conservation through metering and pricing. Supply-side strategies would generally involve increases in storage capacity, desalinization of nonpotable water, or other measures that may require new infrastructure systems and components (Bates et al., 2008).

If the United States is to reduce its greenhouse gas emissions significantly, power and mobility will need to be provided through new methods, technologies, and materials. The reduction of greenhouse gases could potentially also help reduce the impacts of climate change on water resources. Even so, new infrastructure for water systems may be needed to ensure that future supplies are adequate to meet demand. While these challenges are great, continuing to provide water, power, and mobility as was done in the 20th century presents a substantial obstacle to mitigating greenhouse gas emissions and the higher temperatures and extremes of precipitation associated with global climate change.

ENVIRONMENTAL SUSTAINABILITY

For much of the 20th century, relatively little attention was given to the effects of the built environment, including critical infrastructure systems, on the natural environment—oceans, rivers, lakes, ecosystems, raw materials, the air, the soil, and the land. As infrastructure systems were built, much of the country was developed, great swathes of forest and land were cleared, rivers were controlled and channeled, and renewable and nonrenewable natural resources were harvested, extracted, and productively used.

The publication in 1987 of *Our Common Future* by the United Nations' World Commission on Environment and Development, commonly called the Bruntland Commission, called worldwide attention to the issue of sustainable development. *Sustainable development* was defined as follows:

a process of change in which the exploitation of resources, the direction of investments, the orientation of technological development and institutional change are all in harmony and enhance both current and future potential to meet human needs and aspirations. (UN, 1987, Chapter 2)

Environmental sustainability—the regeneration of ecosystems and the judicious use of water, land, and other natural resources now and for the future—has become an urgent need of the 21st century. The processes and materials used to renew existing critical infrastructure systems or to create new components or systems will be significant factors in meeting or failing to meet this imperative.

DISASTER RESILIENCY

Communities and individuals require essential services in order to learn about, react to, and recover from natural or human-made disasters—earthquakes, hurricanes, tornadoes, flooding, terrorism, or accidents. Critical infrastructure systems provide crucial services, including clean water for drinking and for the protection of public health; mobility for the evacuation and repopulation of communities; connectivity for emergency communications and response; and power for hospitals, for safety, security, and incident management, for cooking and refrigerating food, and for the continuity of government operations before, during, and after an event. The condition and performance of these infrastructure systems help determine how effectively a community can react in times of crisis. Critical infrastructure systems that are robust and resilient, as opposed to deteriorating, can also mitigate the effects of a disaster by limiting deaths and injuries, property losses, impacts on ecosystems (for example, uncontrolled discharge of waste), and the time it takes for a community to recover.

In summary, the materials, technologies, and methods chosen to renew critical infrastructure systems will be a determining factor in whether the nation will be able to meet some of the greatest challenges of the 21st century.

3 Underlying Issues

Renewing the nation's critical infrastructure systems to help meet some 21st century imperatives is a radically different task from that of building new systems across undeveloped territory. A comprehensive and coordinated renewal effort must account for a number of underlying issues, including the extensive network of existing systems, their interdependencies, who owns them, how they are financed, and the level of public support for investment.

LEGACY INFRASTRUCTURE

At the end of the 20th century, the United States had 55,000 community drinking water systems; 30,000 wastewater treatment and collection facilities; 4 million miles of roads; 117,000 miles of rail; 11,000 miles of transit lines; 600,000 bridges; 26,000 miles of commercially navigable waterways; 500 train stations; 300 ports; and 19,000 airports (GAO, 2008).

Although infrastructure components and systems are often thought of as "public goods," myriad public- and private-sector organizations are responsible for infrastructure investment, construction, operations, repair, and renewal. Whereas water and wastewater systems are primarily owned and operated by public entities, the private sector owns and operates most power and telecommunications systems. Similarly, state and local authorities are responsible for roads, highways, and bridges, while subways, ports, airports, and railroads are owned and operated by quasi-public or private organizations. Overlaid on these organizations are institutions responsible for developing standards and enforcing compliance with regulations for critical infrastructure systems.

All of these systems and their components have finite lives. Their condition and performance inevitably deteriorate over several decades of use. For their service lives to be extended, these systems require reinvestment through timely maintenance and repair. Eventually they require replacement, in whole or in part.

In 2004 alone, public and private expenditures on critical infrastructure systems totaled $285 billion (Table 3.1). However, these investments have not kept pace with infrastructure needs. The American Society of Civil Engineers, for example, estimates that $2.2 trillion are required over a 5-year period to bring the nation's infrastructure to a good condition that meets the needs of the current population (ASCE, 2009). Studies for the Federal Highway Administration, the Federal Aviation Administration, and other agencies report that about $20 billion more are needed annually to keep transportation services at today's levels—levels that are already inadequate in some areas of the country (CBO, 2008). Another report estimates that the electric utilities industry will need to make a total investment of at least $1.5 trillion between 2010 and 2030 to keep pace with demand (Chupka et al., 2008). The Congressional Budget Office has estimated that an average annual investment of $24.6 billion to $41 billion is needed for drinking water and wastewater systems for the years 2000 through 2019 (CBO, 2002).

Although the needs are great, public investment in infrastructure has declined substantially as a portion of the gross domestic product for the past 50 years (Figure 3.1).

Even before the 2008 financial crisis, the U.S. Government Accountability Office projected that net interest on the national debt, Social Security, Medicare, and Medicaid would consume an increasingly large portion of the federal budget through 2040, limiting the funds available to meet the nation's critical infrastructure challenges (GAO, 2006). Although the 2009 economic stimulus package contains some funding for infrastructure improvements, over the long term the resources available to renew and restructure infrastructure systems and their compo-

TABLE 3.1 Capital Spending on Infrastructure in the United States in 2004, by Category (in Billions of 2004 Dollars)

Infrastructure	Federal Spending	State and Local Spending	Private-Sector Spending	Total
Highways	30.2	36.5	n.a.	66.7
Mass transit	7.6	8.0	n.a.	15.6
Freight railroads	0	0	6.4	6.4
Passenger railroads	0.7	0	0	0.7
Aviation	5.6	6.8	2.0	14.4
Water transportation	0.7	1.7	0.1	2.5
Total transportation	44.8	53.0	8.5	106.3
Water and wastewater	2.6	25.4	n.a.	28.0
Energy (electricity, natural gas, oil pipelines)	1.7	7.7	69.0	78.4
Telecommunications (wired and wireless, Internet service, fiber optics, and broadcasting)	3.9	n.a.	68.6	72.5
TOTAL	53.0	86.1	146.1	285.2

NOTE: n.a., not available.
SOURCE: CBO (2008).

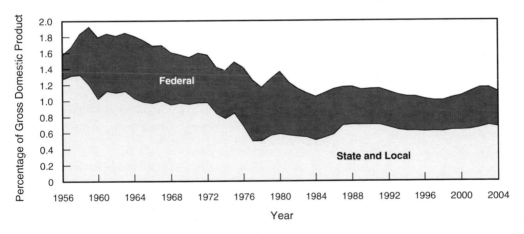

FIGURE 3.1 Public capital spending on transportation and water infrastructure as a percentage of gross domestic product, 1956-2004. NOTE: Includes spending on highways, mass transit, rail, aviation, water transportation, water resources, and water supply and wastewater treatment systems. SOURCE: CBO (2008).

nents will be limited. Efficient use of those funds that are available requires that choices be made about where to invest and about the objectives to be achieved by those investments. At this time, the United States does not have in place a set of objectives, a strategy, policies, or decision-making processes for prioritizing infrastructure investments to meet national objectives. Nor does it have processes or measures for determining the outcomes of investments that are made.

INTERDEPENDENCIES

Infrastructure systems, like environmental corridors, do not stop at community, city, state, or national boundaries. Instead, they physically link regions and markets, crossing jurisdictional and political boundaries. The 2007 water shortage in Atlanta, Georgia, for instance, required negotiations among the three states of Georgia, Florida, and Tennessee for agreement on water flow regulations that affect power plant operation, fishing grounds, and the region's economic activities (Goodman, 2007).

Although critical infrastructure systems were built as stand-alone entities for specific purposes, in actuality they are functionally interdependent. For example, power is needed to treat and pump water, water is needed to cool power and telecommunications equipment or to power steam systems, and telecommunications systems provide automated control for transportation, water, wastewater, and power systems. Many other complex interdependencies exist.

Because these systems share rights-of-way and conduits above- and belowground, they are also geographically interdependent. These functional and geographical interdependencies have resulted in complex systems that regularly interact with one another, sometimes in unexpected and unwelcome ways (Connery, 2008). Because these interdependencies were achieved by default, not by plan, they create vulnerabilities whereby a failure in one system can cascade into other systems, creating more widespread consequences than those resulting from the one system originally experiencing the failure. For example, the failure to repair or replace a deteriorating water main could lead to a break in the main; the flooding of adjacent roads, homes, and businesses; the shutting off of water for drinking and fire suppression; the short-circuiting of underground cables; and the loss of power for a larger community (Figure 3.2). On a much larger scale, the failure of the levees in New Orleans in the aftermath of Hurricane Katrina in 2005 led to the flooding of large portions of the city, knocking out power, water supply, transportation, and wastewater systems for months and even years.

Long-standing institutional arrangements exist with respect to the ownership of, planning for, and building, financing, operating, and regulating of infrastructure systems. Complex proprietary considerations, such as those surrounding the interface between freight and passenger rail (track ownership and other issues), also exist. Such arrangements are often both highly segmented and overlapping, involving some combination of local governments, regional authorities, states, federal regulatory and funding agencies, and private-sector organizations. The current segmented decision-making and governing structure provides few incentives for public- and private-sector groups to discuss crosscutting issues, to collaborate to improve entire infrastructure systems, or to analyze the interdependencies among systems.

FIGURE 3.2 Water main break in Bethesda, Maryland, on December 23, 2008, trapping passengers in cars and creating water and power outages. SOURCE: WTOP Photo/Markette Smith.

OWNERSHIP AND FINANCING STRUCTURES

Today's decision-making and investment strategies, whether public or private, typically focus on one type of infrastructure (e.g., airports), individual components or projects (e.g., a bridge), and the design and construction costs (first costs) of new projects, as opposed to the operation and maintenance costs that will accrue over the 30 to 50 years or more of the infrastructure's service life.

The ownership of a portion of an infrastructure system largely dictates how investments are financed. For example, investments in publicly owned water and wastewater systems are typically funded through federal grants and municipal bonds and thus by taxpayers. Publicly owned systems provide the same level of services to all users, and all users pay the same rates per unit of service. In contrast, fiber-optic systems and towers for telecommunications, television, radio, the Internet, and cellular telephones are built primarily by profit-driven corporations and regulated by public authorities. For-profit businesses typically provide services to those users who are able to pay for them and may offer different levels of service based on willingness to pay.

The differing objectives of owners and operators of infrastructure influence their investment decisions. In general terms, businesses invest in infrastructure and other resources primarily to retain their current customers, expand their customer base, and benefit their stockholders and/or the corporate bottom line. Without some assurance that infrastructure investments can be paid back within a few years, there are few incentives for private-sector firms to make such investments.

Local and state governments, in contrast, must provide services to all households, even if it is not cost-effective to do so. In providing services to all households, governments are also challenged to keep taxes low and contain service costs. Major infrastructure improvements are primarily financed through 15- to 30-year bond programs, which require the support of the local electorate. Faced with a multitude of demands for available funding, including education, health care, and public safety, and reluctant to take on long-term financial obligations, elected officials may decide to defer the maintenance and repair of infrastructure systems indefinitely.

Although most planning, construction, and operation of infrastructure take place at the local, state, or regional level, the influence of the federal government on infrastructure development and management is substantial. This influence is exercised through a multitude of funding programs, standards, and regulations. However, there is no overall concept or set of objectives for critical infrastructure systems, nor is there an integrated federal policy toward infrastructure as a whole to provide a framework within which federal and other infrastructure-related investments might be prioritized and optimized.

Twentieth-century methods for owning and investing in critical infrastructure have resulted in a decision-making environment in which public- and private-sector investments are made on a project-by-project basis. Potential projects for one type of infrastructure are not evaluated against other projects to determine where the greatest overall value might be achieved. The lack of "apples-to-apples comparisons" confounds a prioritization of investments. The segmentation of funding sources among various levels of government and among a multitude of private-sector organizations almost certainly results in the suboptimization of those resources that are invested.

PUBLIC SUPPORT FOR INFRASTRUCTURE INVESTMENT

From the national to the local level, the demands for public services of all kinds exceed available resources. Citizens and jurisdictions are often reluctant to support bonds or other funding for needed infrastructure improvements when other services—police and fire protection, education, health care—are more visible and seem more urgent. In addition, because much of the existing infrastructure is underground or located away from population centers, it engenders an "out-of-sight, out-of-mind" attitude that makes it relatively easy to defer routine maintenance that could prevent failures and extend a system's service life.

Well-publicized cost and schedule overruns in projects like Boston's Central Artery ("Big Dig") (NRC, 2003), coupled with legislative earmarks for projects with unclear objectives—for example, the "bridge to nowhere"[1]—have led many to debate the purpose, value, and costs of infrastructure projects. Typically, it is only when infrastructure systems fail completely that their value is apparent.

The lack of transparency in decision-making processes presents a significant obstacle to building public support for infrastructure investments. Contributing to the lack of transparency is the lack of metrics for quantifying the outcomes of infrastructure investments—for instance, improved efficiency or reliability. Metrics are used by some organizations to measure some aspects of infrastructure investment, such as miles of roads paved or miles of sewer lines repaired. However, such metrics do not help decision makers or the public understand what returns they should expect (i.e., improvements in levels of service) from a given investment in infrastructure.

To date, the public dialogue regarding the use of alternative sources of energy to replace oil and other fossil fuels has not focused on the infrastructure systems and components that will be needed to generate and deliver power from these sources. New systems could potentially have significant environmental and social impacts. If local citizens and officials oppose proposed locations for new facilities and infrastructure, the delays in the siting and construction of required facilities may extend several years or more. Finding ways to deliver mobility and power from alternative energy sources while accounting for local desires is challenging. Finding ways to communicate effectively about what is at stake, as well as the risks, costs, and benefits of differing options, will be essential to building public support.

Tackling the range of issues associated with critical infrastructure renewal is a major challenge in and of itself. Attempting to resolve these issues while also meeting other imperatives of the 21st century is daunting. Meeting such complex challenges requires a new paradigm for critical infrastructure renewal, as outlined in Chapter 4.

---

[1]The "bridge to nowhere" refers to a bridge from Ketchikan, Alaska, on one island in the southeastern part of the state to an airport on another, nearby island. The bridge, proposed for federal funding at a cost of $398 million, became a national symbol of federal "pork barrel" spending. See " 'Bridge to Nowhere' Abandoned." Available at http://www.cnn.com/2007/US/09/22/alaska.bridge.ap. Accessed January 10, 2009.

4 Creating a New Paradigm

Renewing and restructuring an extensive network of existing infrastructure systems to help meet a range of 21st century imperatives constitute a complex, long-term challenge for the United States and its citizens. However, continuing to use the same decision-making processes, construction methods, and operational practices as those used in the 20th century will likely yield the same results: increasing instances of service disruptions, higher operating and repair costs, and the possibility of catastrophic, cascading failures.

Meeting this century's challenges requires a fundamental reexamination of the purposes and value of critical infrastructure systems and of the current processes for infrastructure-related decision making and investment. While daunting, this reexamination can yield a new paradigm from which to develop practical solutions to complex issues.

INGREDIENTS FOR A NEW PARADIGM

Some of the ingredients needed to create the new paradigm referred to above are available today. At the workshop—Toward Sustainable Critical Infrastructure Systems: Framing the Challenges—conducted in May 2008 under the auspices of the National Research Council's Board on Infrastructure and the Constructed Environment, the participants identified an array of new technologies and materials, ongoing initiatives, and financing options that can provide the basis for moving forward with new approaches to infrastructure renewal and investment.

TECHNOLOGIES AND MATERIALS

Research has yielded technologies for monitoring the condition and performance and for improving the management of infrastructure systems; new materials for constructing and repairing infrastructure components; new knowledge about the interrelated nature of water and wastewater, power, transportation, and telecommunications systems; and models to simulate the consequences of deteriorating infrastructure so that preventive actions can be taken to avoid failures.

Today, self-diagnosing, self-healing, and self-repairing systems can be designed to provide for greater resiliency, fewer long-term service disruptions, and lower life-cycle costs (Amin and Stringer, 2008). Buildings can be designed to supply their own electricity using solar collectors and, in some cases, to generate excess electricity that can be sold back to power companies. Wind turbines and more powerful batteries can augment other energy sources for in-place power generation. Decentralized and "package" systems for water and wastewater treatment offer new possibilities for expanding centralized systems or building new stand-alone systems. New telecommunication systems, such as wireless mesh, may be more resilient than current systems and may help allow for more telework and less commuting (Doshi et al., 2007). In Europe magnetic trains are being built to take advantage of existing rail lines while also cutting the cost of energy to power the train by one-third (Toffler Associates, 2008).

Ongoing research has the potential to significantly change how services are provided in the future. For instance, researchers are studying the use of microbial fuel cells to convert complex wastes to electricity, with a focus on wastewater treatment or solid-waste facilities (Lovley, 2009). The first prototype is currently installed on the grounds of a brewery, using organic wastes from the brewery to produce electricity. Research and development of systems that produce electricity and hot water using existing asphalt roadways and parking areas are also underway (WPI, 2008). And water treatment systems that use ultraviolet radiation in place of chlorine, a volatile substance, are being developed (McClean, 2007).

To date, no one has captured the full range of innovative technologies being used or the research that is being undertaken at universities and elsewhere. Some of these existing or emerging technologies may provide breakthroughs that could lead to new ways of providing for essential services and change the nature of today's infrastructure systems. Widespread use of new technologies and materials could create more durable, reliable, and resilient infrastructure with more benign environmental impacts.

ONGOING INITIATIVES

Many groups and individuals have recognized the need to upgrade the nation's critical infrastructure systems and have proposed various solutions.[1] An array of issue papers, plans, and strategies to repair and upgrade infrastructure as a whole or to deal with specific systems, such as water supply systems, have already been developed.[2] At the regional, state, and local levels, community, government, nonprofit, and private-sector organizations are developing and implementing strategies to address one or more infrastructure-related challenges across jurisdictional and political boundaries. The following are examples of such initiatives:

- *For the Memphis metropolitan region,* which includes portions of Tennessee, Mississippi, and Arkansas as well as 40 individual cities,[3] a conceptual framework titled

---

[1]See, for example, "America's Infrastructure: Ramping Up or Crashing Down" (Katz et al., 2007); "Guiding Principles for Strengthening America's Infrastructure" (CSIS, 2006); "A New Bank to Save Our Infrastructure" (Ehrlich and Rohatyn, 2008); "Time for an Infrastructure Overhaul" (Little, 2007); "It's Time to Rebuild America" (Rohatyn and Rudman, 2005); "Report Card for America's Infrastructure—2009" (ASCE, 2009); and "Main Street, Not Wall Street, Should Fix Crumbling Infrastructure" (Sebelius and Stern, 2008).

[2]See, for example, *Critical Issues in Transportation* (TRB, 2006), *The Clean Water and Drinking Water Infrastructure Gap Analysis* (U.S. EPA, 2002); *An Economic Strategy for Investing in America's Infrastructure* (Deshpande and Elmendorf, 2008); *Energy and Water Distribution Interdependency Issues: Best Practices and Lessons Learned* (U.S. EPA, 2005); *Future Investment in Drinking Water and Wastewater Infrastructure* (CBO, 2002).

[3]The area is described as follows: Although the Memphis metropolitan region is composed of two distinct and geographically unequal parts due to the effect of the Mississippi River, it is still one unit. Two automobile and two rail bridges connect the city across the Mississippi. While the urbanized areas

"A Strategic Global Future" has been issued with the stated mission: "Firmly establish the Memphis region as a dynamic, growing, energetic, metropolitan region strongly connected to the global economy" (Boyle and Associates, 2009). The framework addresses many aspects of the region's economy, its critical infrastructure systems, and a plan for investment and management that explicitly crosses jurisdictions to enable significant changes.

- *In the state of Indiana,* the Departments of Transportation and Agriculture have undertaken "a joint initiative to explore the potential for major statewide infrastructure improvements that can strategically support and drive Indiana's economic growth" (Purdue University, 2006, p. 2). The Indiana State "Pipe Dream" Workshop was held to "identify new and dramatic improvements in underground transportation infrastructure that would accommodate current and future needs for . . . Energy, Broadband, Communications, Livestock/Agriculture, Storm/Wastewater Treatment, and Fresh Water Supply" (ibid).

- *The America 2050 coalition* is a "national initiative to meet the infrastructure, economic development and environmental challenges of the nation as we prepare to add about 130 million additional Americans by the year 2050" (America 2050, 2009). The goal of this effort, guided by a coalition of regional planners, scholars, and policy makers, is to develop a framework for the nation's future growth that considers trends such as rapid population growth and demographic change, global climate change, the rise in foreign trade, and infrastructure systems that are reaching capacity.

- *Blueprint America* was developed by nonprofit foundations to "shine an unyielding spotlight" on America's decaying and neglected infrastructure and to educate the public about critical infrastructure-related issues and the policy choices that will need to be made (PBS, 2009).

---

on the two sides of the river are not contiguous, they are highly interactive and have developed closely together. They are therefore considered to be part of the same metropolitan area (Boyle and Associates, 2009).

Existing issue papers and plans provide a wealth of detailed information about infrastructure issues and needs. Ongoing, community-based initiatives and strategies provide a source of creative and collaborative approaches for infrastructure renewal that cross jurisdictional and institutional boundaries. In the absence of an overarching vision or concept for critical infrastructure systems, these plans, initiatives, and strategies have been independently developed, each with its own set of objectives.

FINANCE MECHANISMS

Faced with expensive infrastructure renewal projects, individuals, organizations, and communities have proposed innovative financing options that include public-private partnerships (Orr, 2007), restructuring of the Highway Trust Fund, increased reliance on user fees such as those for toll roads and high-occupancy toll lanes, use of public pension funds (Sebelius and Stern, 2008), privatization (Anderson, 2008), and the establishment of a national infrastructure bank (Ehrlich and Rohatyn, 2008).

Ownership and financing mechanisms entail differing levels of risk, involve issues of social equity (for example, access to services according to one's ability to pay), and impact how the public values the services provided. Each approach has short- and long-term social and financial implications, and no single approach would be appropriate to all situations. Choosing the best approach for a specific situation would depend on the objectives to be achieved.

THE NEED FOR A FRAMEWORK

Although technological advances, community-based initiatives, and financing options offer the promise of new ways to approach critical infrastructure renewal, they have been ad hoc, often focusing on one issue, one type of system, or one set of solutions. By concentrating on single projects, technologies, financing mechanisms, or narrowly defined objectives, ad hoc efforts run the risk of wasting scarce resources and increase the probability of serious, unintended consequences. A framework is needed to create a structure within which ongoing activities, knowledge, and technologies can be aligned and leveraged to support critical infrastructure renewal and also to help achieve some of the nation's 21st century imperatives.

A FRAMEWORK FOR DEVELOPING SUSTAINABLE CRITICAL INFRASTRUCTURE SYSTEMS

The purpose of a framework is to create an environment for developing short- and long-term solutions to complex issues involving a multitude of stakeholders. A framework can provide the structure for establishing public expectations about the reliability, resiliency, efficiency, and cost of critical infrastructure systems and can guide actions for solutions that are physically resilient, socially equitable, cost-effective, and environmentally viable. The following are the essential components of such a framework:

- *A broad and compelling vision* that will inspire individuals and organizations to pull together to help meet 21st century imperatives by renewing the nation's critical infrastructure systems. Such a vision would focus on a future of economic competitiveness, energy independence, environmental sustainability, and quality of life, not a legacy of deteriorating concrete, steel, and cables.
- *A focus on providing the essential services involving water and wastewater, power, mobility, and connectivity*—in contrast to upgrading individual physical facilities—to foster innovative thinking and solutions.
- *Recognition of the interdependencies among critical infrastructure systems* to enable the achievement of multiple objectives and to avoid narrowly focused solutions that may well have serious, unintended consequences.
- *Collaborative, systems-based approaches* to leverage available resources and provide for cost-effective solutions across institutional and jurisdictional boundaries.
- *Performance measures* to provide for greater transparency in decision making by quantifying the links among infrastructure investments, the availability of essential services, and other national imperatives.

VISION

Americans have undertaken great challenges when leaders have effectively communicated the importance and significance of the issues at stake. President Dwight D. Eisenhower presented a vision for national security based in part on the construction of an interstate highway system. His vision inspired the country's public- and private-sector leaders to design, plan, finance, and build that system. President John F. Kennedy's call for the nation to land a man on the Moon and return safely to Earth within the decade of the 1960s was similarly inspiring. This challenge was met through a collaborative effort of scientists, engineers, businessmen, and government officials and was supported by taxpayers.

The United States does not currently have a vision for its critical infrastructure systems to guide the development of concepts, strategies, stated objectives for their future configuration, level of performance, or level of services. Current policies, procedures, and decision-making processes are unlikely to result in the development of a network of systems that will meet current and future social, economic, and environmental needs.

The last congressionally initiated review of the condition of the nation's critical infrastructure systems concluded in 1988 with the publication of *Fragile Foundations: A Report on America's Public Works*. The report found "convincing evidence that the quality of America's infrastructure is barely adequate to fulfill current requirements and insufficient to meet the demands of future economic growth and development" (NCPWI, 1988, p. 1). To upgrade infrastructure systems, *Fragile Foundations* recommended a broad-based, long-term strategy involving the government, the private sector, and the public. Twenty years later, government and industry leaders have not called for a broad-based, collaborative, long-term strategy, and consequently it has not been developed.

A vision is now needed that will inspire and rally business, community, academic, and government leaders to tackle issues related to critical infrastructure systems. It should not be a vision of concrete, steel, and cables, but rather one of expectations for economic competitiveness, reduced dependence on imported oil, a high quality of life, and harmony with the environment. In the absence of such a vision, ad hoc initiatives and investments for critical infrastructure systems driven by economic forces or disaster recoveries will continue, but it will be difficult to integrate these into a coherent approach to meet 21st century needs.

FOCUS ON PROVIDING ESSENTIAL SERVICES

Citizens and businesses expect that essential services—water and wastewater, power, mobility, and connectivity—will be available without interruptions. However, business and population growth have already outpaced the capacity of existing systems to meet those expectations, as evidenced by transportation congestion, air and water pollution, and increasing instances of power and service disruptions.

If stakeholders are to understand fully what is at risk and what choices need to be made, the public dialogue needs to be recast as a discussion on how best to provide essential services—as opposed to its current focus on the merits and deficiencies of individual physical systems. As part of this discussion, it will be necessary to develop answers to questions such as the following:

- What are the public's expectations for the levels of services to be provided by critical infrastructure systems?
- What are their expectations with respect to the resiliency of these systems?
- What actions will be necessary to achieve those expectations?
- How much money are people and businesses willing to invest now and in the coming years?
- What alternatives are available to traditional practices for providing essential services?
- What actions are needed to develop systems that are physically, socially, financially, and environmentally sustainable?

Shifting the conversation from deteriorating, engineered systems to the provision of essential services that affect everyone's quality of life can refocus stakeholders' attention on the ultimate value of such services. Doing so will provide opportunities for more creative thinking, greater engagement by a wider range of stakeholders, and a more robust and diverse array of possible solutions for service delivery and infrastructure renewal.

RECOGNITION OF INTERDEPENDENCIES AMONG CRITICAL INFRASTRUCTURE SYSTEMS

Providing for resilient and reliable infrastructure systems requires crosscutting, collaborative approaches to enable the identification and mitigation of vulnerabilities and the leveraging of resources and solutions. Such approaches are possible and practical, as evidenced by the efforts highlighted earlier in this chapter that are occurring in the Memphis metropolitan region, in Indiana, and in other places.

Finding ways to collaborate across institutional and jurisdictional boundaries can help to enable the achievement of multiple objectives, such as reducing greenhouse gas emissions, protecting water supplies, and working toward environmental sustainability. In addition, it can help to avoid narrowly focused solutions with serious, unintended consequences by bringing more information and more stakeholders to the table. By considering the interactions of water, wastewater, power, transportation, telecommunications, and the environment, it should be possible to develop solutions that meet multiple objectives and are sustainable for future generations.

COLLABORATIVE, SYSTEMS-BASED APPROACHES

Existing institutional arrangements and decision-making processes inhibit effective thinking about the interactions among various infrastructure systems, about their overall performance in delivering services, and about the costs of operating and maintaining these systems over a 50- to 100-year life span. Shifting the public dialogue to focus on essential services, the regional nature of infrastructure systems, and their interdependencies will provide opportunities to bring together stakeholders from a range of infrastructure-related organizations to discuss issues that cut across institutional, jurisdictional, and political boundaries. In doing so, they can potentially identify new ways to leverage resources, to optimize investments, and to identify solutions that meet multiple objectives.

PERFORMANCE MEASURES

Although infrastructure systems are built and operated to provide essential, complex, and varied services for societies, their performance or effectiveness is seldom evaluated against social objectives, such as health and safety, cost-effectiveness, or reliability (NRC, 1995). The lack of performance measures inhibits transparency and effective decision making about infrastructure-related investments because it is not clear what results can be expected or what results are actually achieved by such investments.

A first step in developing an effective performance measurement system is to establish goals and objectives for the elements to be measured—for example, the level of services to be provided by critical infrastructure systems. Data on current levels of service can provide a baseline. Investments in infrastructure can then be measured against the baseline to determine if the levels of services are improving or declining, allowing appropriate actions to be taken.

At least three broad categories of measures will be needed to evaluate the performance of infrastructure: effectiveness, reliability, and cost (NRC, 1995). A variety of ways could be used to provide real-time performance data and public feedback, including sensors and other monitoring technologies.

Readily available, transparent performance information could lead to changes in the behaviors of institutions and individuals, alter perceptions about the value of infrastructure, and lead to greater accountability regarding the results of infrastructure investments. For example, the first net-zero electric commercial building[4] in the United States has a dynamic graphic display in the lobby that shows the "real time" operation of the building in a dashboard format and has inspired employees to save energy by turning off lights and taking other similar measures (Grabowski, 2008).

A multiple-objective performance measurement system would promote greater transparency in decision making, improved information for making decisions, and a better understanding about the links between infrastructure investment and economic competitiveness, quality of life, and environmental quality. It could also help to communicate what is at stake and the risks involved when critical infrastructure investments are being considered. Performance measures may also help to establish long-term public support for investments in infrastructure among its users—citizens, businesses, nonprofit organizations,

---

[4]Defined as a building with a net energy consumption of zero over a typical year: that is, energy produced minus energy used equals zero (Grabowski, 2008).

governments, and other public- and private-sector institutions (NRC, 1995).

MOVING FORWARD

An important first step in creating a new paradigm is to bring together those who have an essential stake in meeting 21st century imperatives and who are already involved in sustainable infrastructure efforts. They include infrastructure owners, designers, engineers, financiers, regulators, and policy makers, as well as ecologists, community activists, scientists, and researchers. Working within the framework, experts in such areas could begin to identify a full range of new approaches, technologies, and materials for providing the services of mobility, connectivity, water, wastewater, and power to meet multiple objectives. They could also identify new approaches to decision making, finance, operations, and processes related to infrastructure. The results of such a gathering could serve to initiate a longer-term, collaborative effort to develop a vision, concepts, and objectives for the nation's critical infrastructure systems and then to identify the policies, practices, and resources required to implement the vision. The results could be critical infrastructure systems that are physically, economically, socially, and environmentally sustainable for the next 50 years.

References

America 2050. 2009. America 2050. Information available at http://www.america2050.org/about.html. Accessed January 12, 2009.

Amin, S.M. 2008. For the good of the grid. IEEE Power and Energy Magazine. November/December: pp. 48-59.

Amin, M., and J. Stringer. 2008. The electric power grid: Today and tomorrow. MRS Bulletin. Vol. 33. Available at www.mrs.org/bulletin. Accessed December 12, 2008.

Anderson, J. 2008. Cities debate privatizing public infrastructure. New York Times. August 27. Available at http://www.nytimes.com/2008/08/27/business/27fund.html?partner=rssnyt. Accessed December 12, 2008.

ASCE (American Society of Civil Engineers). 2005. Report Card for America's Infrastructure. Available at http://www.asce.org/reportcard/2005/index.cfm. Accessed July 6, 2008.

ASCE. 2007. Raising the grades: Small steps for big improvements in America's failing infrastructure—Action plan for the 110th Congress. Available at http://www.asce.org/reportcard/2005/actionplan07.cfm. Accessed December 12, 2008.

ASCE. 2009. 2009 Report Card for America's Infrastructure. Available at http://www.asce.org/reportcard/2009/. Accessed January 30, 2009.

Bates, B.C., A.W. Kundzewicz, S. Wu, and J.P. Palutikof, eds. 2008. Climate Change and Water. Technical Paper of the Intergovernmental Panel on Climate Change, IPCC Secretariat, Geneva, Switzerland

Belson, K., and A. DePalma. 2007. Asbestos and aging pipes remain buried hazards. New York Times. July 19. Available at http://www.nytimes.com/2007/07/19/nyregion/19steam.html?_r=1. Accessed December 12, 2008.

Beven, J.L., L.A. Avila, E.S. Blake, D.P. Brown, J.L. Franklin, R.D. Knabb, R.J. Pasch, J.R. Rhome, and S.R. Stewart. 2007. Annual Summary: Atlantic Hurricane Season of 2005. Available at http://www.aoml.noaa.gov/general/lib/lib1/nhclib/mwreviews/2005.pdf. Accessed February 6, 2009.

Boyle, S., and Associates. 2009. The Memphis Sourcebook. Available at http://www.memphisregion.com/future.asp. Accessed January 9, 2009.

CBO (Congressional Budget Office). 2002. Future Investment in Drinking Water and Wastewater Infrastructure. Washington, D.C.

CBO. 2008. Investing in Infrastructure: Testimony. Statement of Peter R. Orszag, Director, before the Committee on Finance, United States Senate. July 10, 2008. Available at http://www.cbo.gov/ftpdocs/95xx/doc9534/MainText.1.2.shtml. Accessed November 30, 2008.

Chupka, M.W., R. Earle, P. Fox-Penner, and R. Hledik. 2008. Transforming America's Power Industry: The Investment Challenge 2010-2030. Available at http://www.eei.org. Accessed December 12, 2008.

Connery, N.R. 2008. Development of next-generation U.S. infrastructure systems: A framework for national policy. Public Works Management and Policy. Vol. 12, No. 3: pp. 479-482. Sage Publications: Los Angeles.

CSIS (Center for Strategic and International Studies). 2006. Guiding Principles for Strengthening America's Infrastructure. Available at http://www.csis.org/media/csis/pubs/060327_infrastructure_principles.pdf. Accessed December 12, 2008.

Deshpande, M., and D.W. Elmendorf. 2008. An Economic Strategy for Investing in America's Infrastructure. The Hamilton Project: Brookings Institution. Available at http://www.brookings.edu/papers/2008/07_infrastructure_elmendorf.aspx. Accessed December 12, 2008.

DHS (Department of Homeland Security). 2009. National Infrastructure Protection Plan: Partnering to Enhance Protection and Resiliency. Washington, D.C.: DHS.

Doshi, V., G. Schulman, and D. Gabaldon. 2007. Lights! Water! Motion! In Global Perspective, Strategy+Business Issue. Washington, D.C.: Booz, Allen, Hamilton.

EC (European Commission). 1999. ESDP (European Spatial Development Perspective): Towards Balanced and Sustainable Development of the Territory of the European Union. Available at http://ec.europa.eu/regional_policy/sources/docoffic/official/reports/pdf/sum_en.pdf. Accessed February 6, 2009.

Ehrlich, E., and F.G. Rohatyn. 2008. A new bank to save our infrastructure. New York Review of Books. Vol. 55, No. 5. Available at www.nybooks.com/articles/21873. Accessed December 12, 2008.

EIA (Energy Information Administration). 2008a. Energy in Brief: What Everyone Should Know About Energy. Available at http://tonto.eia.doe.gov/energy_in_brief/foreign_oil_dependence.cfm. Accessed December 12, 2008.

EIA. 2008b. Emissions of Greenhouse Gases Report. Release date December 3, 2008. Available at http://www.eia.doe.gov/oiaf/1605/ggrpt/index.html?featureclicked=2&. Accessed December 12, 2008.

Gallis, Michael. 2008. Globalization and Infrastructure Needs. Presentation at Workshop: Toward Sustainable Critical Infrastructure Systems: Framing the Challenges. May 7-8, 2008. Washington, D.C.

GAO (Government Accountability Office). 2006. The Nation's Long-Term Fiscal Outlook. Washington, D.C. Available at www.gao.gov. Accessed September 13, 2008.

GAO. 2008. Physical Infrastructure: Challenges and Investment Options for the Nation's Infrastructure. GAO-08-763T. Washington, D.C. Available at www.gao.gov. Accessed July 14, 2008.

Goodman, B. 2007. U.S. acts to bolster supply of water for Atlanta. New York Times. November 17. Available at http://www.nytimes.com/2007/11/17/us/17water.html?_r=1&ref=us&pagewanted=print. Accessed December 12, 2008.

Grabowski, J. 2008. First Net Zero Electric Commercial Building in the U.S.: 31 Tannery Project. Available at http://www.ibtta.org/files/PDFs/Grabowski_John.pdf. Accessed February 12, 2009.

Gramlich, E.M. 1994. Infrastructure investment: A review essay. Journal of Economic Literature. Vol. 32, No. 3: pp. 1176-1196.

Kates, R.W., C.E. Colten, S. Laska, and S.P. Leatherman. 2006. Reconstruction of New Orleans after Hurricane Katrina: A research perspective. Proceedings of the National Academy of Sciences of the United States of America. Available at http://belfercenter.ksg.harvard.edu/files/xstandard/kates_pnas_katrina_2006.pdf. Accessed February 6, 2009.

Katz, B., R. Puentes, and C. Geissler. 2007. America's Infrastructure: Ramping Up or Crashing Down. Available at http://www.brookings.edu/~/media/Files/rc/papers/2008/01_infrastructure_katz_ puentes/01_infrastructure_katz_puentes.pdf. Accessed January 12, 2009.

Little, R.G. 2007. Time for an infrastructure overhaul. In Dialogue Rebuilding, Urban Land. Pp. 40-42. Washington, D.C.: Urban Land Institute.

Lovley, D.R. 2009. Future shock from the microbe electric. Microbial Biotechnology. Vol. 2, pp. 139-141.

McClean, J. 2007. Using UV for dechlorination. Water and Wastes Digest. 2007. Vol. 47, No. 11. Available at http://www.wwdmag.com/Using-UV-for-dechlorination-article8635. Accessed February 12, 2009.

Minkel, J.R. 2008. The 2003 Northeast Blackout—Five years later. Scientific American. Available at http://www.sciam.com/article.cfm?id=2003-blackout-five-years-later&print=true. Accessed December 12, 2008.

MnDOT (Minnesota Department of Transportation). 2007. Mn/DOT announces apparent bid winner for I-35W bridge project. September 19. Available at www.dot.state.mn.us/newsrels/07/09/19-35wresults.html. Accessed December 12, 2008.

Mongelluzzo, B. 2008. Biggest ships a quandary for East Coast port plans. Journal of Commerce Online. Available at http://www.joc.com/articles/Printable.asp?sid=47076. Accessed February 12, 2009.

MSTPRC (Ministry of Science and Technology of the People's Republic of China). 2006. National High-Tech R&D Program (863 Program). Available at http://www.most.gov.cn/eng/programmes1/200610/t20061009_36225.htm. Accessed February 6, 2009.

NCPWI (National Council on Public Works Improvement). 1988. Fragile Foundations: A Report on America's Public Works: Final Report to the President and Congress. Washington, D.C.

NRC (National Research Council). 1995. Measuring and Improving Infrastructure Performance. Washington, D.C.: National Academy Press.

NRC. 2003. Completing the "Big Dig": Managing the Final Stages of Boston's Central Artery/Tunnel Project. Washington, D.C.: The National Academies Press.

NRC. 2008a. The National Academies Summit on America's Energy Future: Summary of a Meeting. Washington, D.C.: The National Academies Press.

NRC. 2008b. Potential Impacts of Climate Change on U.S. Transportation. TRB Special Report 290. Washington, D.C.: The National Academies Press.

NRC. 2008c. Water Implications of Biofuels Production in the United States. Washington, D.C.: The National Academies Press.

PBS (Public Broadcasting Service). 2009. Blueprint America. Available at http://www.pbs.org/wnet/blueprintamerica/. Accessed January 10, 2009.

Purdue University. 2006. Proceedings of the Indiana State "Pipe Dream" Workshop: July 18, 2006. Available at http://rebar.ecn.purdue.edu/jtrp/Activities/7-18-06Proceedings_Ver_1.2.pdf. Accessed February 6, 2009.

Orr, R.J. 2007. The rise of infra funds. Global Infrastructure Report 2007. Project Finance International. Available at http://crgp.stanford.edu/publications/articles_presentations/Orr_01_Infra_funds_ 2007pfie.pdf. Accessed November 13, 2008.

Rohatyn, F.G., and W. Rudman. 2005. It's time to rebuild America. Washington Post.com. December 13. Available at http://www.washingtonpost.com/wp-dyn/content/article/2005/12/12/AR2005121201263.html. Accessed September 30, 2008.

Sebelius, K., and A. Stern. 2008. Main Street, not Wall Street, should fix crumbling infrastructure. Christian Science Monitor. May 7.

Toffler Associates. 2008. Defining the Future of Sustainable Infrastructure. Available at www.toffler.com. Accessed November 17, 2008.

TRB (Transportation Research Board). 2006. Critical Issues in Transportation. Washington, D.C.: The National Academies Press.

TTI (Texas Transportation Institute). 2005. 2005 Urban Mobility Study. College Station, Texas. Available at http://mobility.tamu.edu/ums/. Accessed October 13, 2008.

UN (United Nations). 1987. Our Common Future. Available at http://www.un-documents.net/ocf-02.htm#I. Accessed February 6, 2009.

U.S. DOC (U.S. Department of Commerce). 2008. Statistical Abstract of the United States: 2008. Washington, D.C.

U.S. EPA (U.S. Environmental Protection Agency). 2002. The Clean Water and Drinking Water Infrastructure Gap Analysis. Office of Water. EPA-816-R-02-020. Washington, D.C.

U.S. EPA. 2005. Energy and Water Distribution Interdependency Issues: Best Practices and Lessons Learned. Summary Report of 2005 Energy and Water Distribution Exercise. EPA 600/R-07/042. Washington, D.C.

WPI (Worcester Polytechnic Institute). 2008. Alternative Energy Hits the Road: Research at WPI Explores Turning Highways and Parking Lots into Solar Collectors. Available at http://www.wpi.edu/News/Releases/20089/asphaltnews.html. Accessed February 6, 2009.

Appendixes

A Biosketches of Committee Members

David J. Nash, *Chair,* has more than four decades of experience in building, design, and program management for both the U.S. Navy and the private sector. His experience includes the management of multibillion-dollar physical asset programs, including the U.S. Navy's shore installations worldwide and the reconstruction of Iraq's infrastructure. He is the president of Dave Nash and Associates, LLC, a company focused on project development and execution in emerging markets and in the United States. He was previously the president of government operations for the international engineering firm of BE&K, Inc., headquartered in Birmingham, Alabama.

Prior to joining BE&K, Admiral Nash served as director of the Iraq Program Management Office under the Coalition Provisional Authority and, later, as director of the Iraq Reconstruction Management Office under the U.S. Department of State. Under his direction, these organizations managed the $18.4 billion Iraq infrastructure reconstruction program. His service in the U.S. Navy spanned 33 years, during which he served as an officer in the Civil Engineer Corps (CEC). Among his many leadership positions, he served at the top of the CEC as commander, Naval Facilities Engineering Command, and chief of civil engineers.

E. Sarah Slaughter, *Vice Chair,* is a National Academies' associate who currently conducts a laboratory on sustainable development at the Sloan School of Business at the Massachusetts Institute of Technology (MIT). Dr. Slaughter previously served as the founder and president of MOCA Systems, Inc., a provider of simulation systems for managing capital projects. Before founding MOCA in 1999, she was a professor in the Department of Civil and Environmental Engineering at MIT specializing in construction management. Prior to joining MIT, she was a professor of civil and environmental engineering at Lehigh University and a researcher at the Center for Advanced Technology for Large Structural Systems. She has researched innovations in design and construction for more than 20 years, and has published more than 50 articles and books on this topic. Dr. Slaughter is a recognized leader in her field and has been selected for several prominent committees and awards. She received all of her degrees from MIT, including a BS in civil engineering, an MS in civil engineering and technology and policy, and a PhD in civil engineering and management science.

Massoud Amin is a professor of electrical and computer engineering, holds the Honeywell/H.W. Sweatt Chair in Technological Leadership, and is the director of the Center for the Development of Technological Leadership at the University of Minnesota. In addition to his administrative responsibilities, he serves as the director of graduate studies for the Management of Technology Program and teaches several courses.

Prior to joining the University of Minnesota in March 2003, Dr. Amin was with the Electric Power Research Institute (EPRI) in Palo Alto, California. In the aftermath of the tragic events of September 11, 2001, he directed all security-related research and development at EPRI, including the Infrastructure Security Initiative and the Enterprise Information Security. Before October 2001, he served as manager of mathematics and information science at EPRI, where he led strategic research in modeling, simulation, optimization, and adaptive control of national infrastructures for energy, telecommunication, transportation, and finance. He served as a member of the National Research Council's (NRC's) Board on Infrastructure and the Constructed Environment from 2001 to 2007 and is currently a member of the Board on Mathematical Sciences and Applications. Dr. Amin also serves on the NRC Committee on Enhancing the Robustness and Resilience of Future Electrical Transmission and Distribution in the United States to Terrorist Attack.

Mark Eberhart is a professor of chemistry and materials science with the Colorado School of Mines, where he is actively involved in communicating the concepts of science and engineering to the public. He has published two books of popular science: *Why Things Break: Understanding the World by the Way It Comes Apart* (Harmony Books, 2003) and *Feeding the Fire: The Lost History and Uncertain Future of Mankind's Energy Addiction* (Harmony Books, 2007). Dr. Eberhart earned his PhD in materials science and engineering from MIT in 1983. He is currently a consultant to the Public Broadcasting Service's program NOVA in its development of a series on materials science. He was the American Chemical Society's Diplomacy Fellow in 2004-2005.

Henry J. Hatch retired from the U.S. Army Corps of Engineers, where he served as chief of engineers and as commander. He was elected to the National Academy of Engineering in 1992. His interests include all aspects of civil engineering; public policy related to infrastructure, the environment, and sustainable development; military engineering; leadership; engineering management; and water resources development. General Hatch was the chair of the Board on Infrastructure and the Constructed Environment from 2005 to 2007.

General Hatch was the chief operating officer of the American Society of Civil Engineers (ASCE) from 1997 to 1999. Before joining ASCE, he was president and chief executive officer of Fluor Daniel Hanford, Inc., where he directed a $5 billion, 5-year management contract for the U.S. Department of Energy's environmental cleanup at the Hanford Nuclear Site. General Hatch is a registered professional engineer in the District of Columbia and a member and past national president of the Society of American Military Engineers. He graduated from the U.S. Military Academy at West Point and has an MS in geodetic science from Ohio State University.

Sue McNeil is a professor of civil and environmental engineering and urban affairs and public policy at the University of Delaware. She was formerly director of the Urban Transportation Center and professor in the College of Urban Planning and Public Affairs and the Department of Civil and Materials Engineering at the University of Illinois at Chicago (UIC). Prior to joining UIC, she was a professor of civil and environmental engineering and of engineering and public policy at Carnegie Mellon University.

She is an expert in transportation infrastructure management with emphasis on the application of advanced technologies, economic analysis, analytical methods, and computer applications. Dr. McNeil is a member of the executive committee of the Transportation Research Board (TRB) and served on the Board on Infrastructure and the Constructed Environment from 2001 to 2007. She chairs the TRB Committee on Asset Management. She chaired the ASCE Urban Transportation Division Committee on Transportation Facilities Management (1988-1993) and is a founding associate editor of the ASCE *Journal of Infrastructure Systems*. Dr. McNeil is a registered professional engineer.

Robert Prieto is senior vice president of the Fluor Corporation's Industrial and Infrastructure Group. Fluor Corporation provides services on a global basis in the fields of engineering, procurement, construction, operations, maintenance, and project management. Mr. Prieto has extensive experience in developing world-class projects for the global infrastructure industry and has participated internationally on task forces and forums focused on delivering critical infrastructure to meet the growing needs of cities and of homeland security. He is the author of *Strategic Program Management,* which addresses some of the major strategic issues that affect all large construction programs and projects, including changes in program governance structures.

Previously Mr. Prieto served as one of three presidential appointees to the Asia Pacific Economic Cooperation Business Advisory Council, as well as serving on the boards of several international, industrial, and educational organizations. He co-chaired the New York City Partnership's Infrastructure Task Force, established following September 11, 2001, and chaired the historic meeting of the World Economic Forum's Engineering and Construction Governors held in New York City in February 2002. Under his leadership, the Disaster Response Network of the World Economic Forum was initiated. Mr. Prieto holds a bachelor's degree from New York University and a master's degree from the Polytechnic Institute of New York, where he currently serves as a trustee.

Garret P. Westerhoff is chairman emeritus of Malcolm Pirnie, Inc., and a leading national expert on water treatment systems, water resources planning, and the design of drinking water projects. A pioneer of innovative technologies, he led design of the nation's first major installation of granular-activated carbon treatment to remove a broad spectrum of organic contaminants from drinking water. Mr. Westerhoff was elected to the National Academy of Engineering in 2000 for leadership in the application of new technologies for drinking water treatment and for international contributions to utility management. He is a member of the NRC's Water Science and Technology Board.

A licensed professional engineer in 14 states and a professional planner in New Jersey, Mr. Westerhoff holds BS and MS degrees in engineering from the New Jersey Institute of Technology, is an honorary member of the American Water Works Association, and is actively involved in leadership roles in numerous professional societies. He is a prolific author with numerous published articles, and he has authored two major texts for drinking water utilities, *The Changing Water Utility: Creative Approaches to Effectiveness and Efficiency* (1998) and *The Evolving Water Utility: Pathways to Higher Performance* (2003), both published by the American Water Works Association.

B Workshop Participants

Brenda Bohlke, Myers Bohlke Enterprises, LLC, Great Falls, Virginia

Jack Buffington, Mack-Blackwell Center for Rural Transportation, University of Arkansas-Little Rock

George Bugliarello, Polytechnic Institute of New York University

Nancy Rutledge Connery, Independent Consultant, Woolwich, Maine

Wayne Crew, Construction Industry Institute, Austin, Texas

Ivan Damnjanovic, Texas A&M University, College Station

Jesus de la Garza, Virginia Tech, Blacksburg

Reginald DesRoches, Georgia Institute of Technology, Atlanta

John Dismukes, University of Toledo, Toledo, Ohio

Terrel Dorn, Government Accountability Office, Washington, D.C.

Dennis Dunne, dddunne & associates, Scottsdale, Arizona

Beverly Dyer, U.S. Department of Energy, Washington, D.C.

Alejandro Fernandez, Department of Homeland Security, Washington, D.C.

Paul Fisette, University of Massachusetts, Amherst

Michael Gallis, Michael Gallis & Associates, Charlotte, North Carolina

Michael Garvin, Virginia Tech, Blacksburg

Paul Gilbert, Parsons Brinckerhoff, Quade and Douglas, Seattle, Washington

Russell Gwatney, Gwatney Companies, Memphis, Tennessee
Alan Haggerty, Office of the Secretary of Defense,
    Washington, D.C.
Yacov Haimes, University of Virginia, Charlottesville
Sheila Hollis, Duane Morris LLC, Washington, D.C.
Mary Ellen Hynes, Department of Homeland Security,
    Washington, D.C.
Adam Krantz, The National Association of Clean Water
    Agencies, Washington, D.C.
Cynthia Lane, American Water Works Association,
    Washington, D.C.
Andrew Lemer, Transportation Research Board,
    Washington, D.C.
Richard Little, Keston Institute for Public Finance and
    Infrastructure Policy, Los Angeles, California
Peter Marshall, Dewberry Company, Norfolk, Virginia
Mike Meyer, Georgia Institute of Technology, Atlanta
Paul Mlakar, U.S. Army Engineer Research and Development
    Center, Vicksburg, Mississippi
Get Moy, DMJM Management, Arlington, Virginia
Pamela Murray-Tuite, Virginia Tech, Blacksburg
Yuko Nakanishi, Nakanishi Research and Consulting, LLC,
    New York City
Priscilla Nelson, New Jersey Institute of Technology, Newark
Richard Norment, National Council for Public-Private
    Partnerships, Washington, D.C.
Mark Palmer, National Institute of Standards and Technology,
    Gaithersburg, Maryland
Stephan Parker, Transportation Research Board,
    Washington, D.C.
Lewis Perelman, Management/Policy Consultant and Analyst,
    Woodbridge, Virginia
James B. Porter, E.I. du Pont de Nemours and Company,
    Wilmington, Delaware
Susan Hill Skemp, Florida Atlantic University, Dania Beach
David Skiven, General Motors Worldwide Facilities Group,
    Detroit, Michigan
Dimitra Syriopoulou, U.S. Army Corps of Engineers,
    Washington, D.C.
Catherine Tehan, American Society of Civil Engineers,
    Washington, D.C.
Hans Van Winkle, Hill International, Inc., Marlton, New Jersey

William "Al" Wallace, Rensselaer Polytechnic Institute, Troy, New York

James A. Wilding, Washington Metropolitan Airports Authority, Fairfax, Virginia

Richard Wright, Practice, Education and Research for Sustainable Infrastructure (PERSI), ASCE, Montgomery Village, Maryland

C Workshop Agenda and Draft Infrastructure Challenges

AGENDA: TOWARD SUSTAINABLE CRITICAL INFRASTRUCTURE SYSTEMS: FRAMING THE CHALLENGES WORKSHOP

MAY 7 AND 8, 2008
NATIONAL ACADEMY OF SCIENCES
2101 CONSTITUTION AVENUE, N.W.
WASHINGTON, D.C.

WORKSHOP OBJECTIVES

- Frame fundamental challenges in moving toward critical infrastructure systems (water, wastewater, transportation, communications, power, telecommunications) that are physically, socially, economically, and environmentally sustainable.
- Identify a range of policies, strategies, technologies, processes, and other lines of inquiry with the potential to address these challenges.

WEDNESDAY, MAY 7, 2008

8:15 a.m.   Welcome and Purpose of the Workshop and
               Desired Outcomes
               David J. Nash, *Chair,* and E. Sarah Slaughter,
               Vice Chair

8:30 a.m.   Globalization and Infrastructure Needs
               Michael Gallis, Michael Gallis and Associates

9:15 a.m.   The Next-Generation U.S. Infrastructure
               Nancy Rutledge Connery, Consultant

10:00 a.m.   Break

10:30 a.m.   Presentations of the Draft Challenges Objectives
               and Ground Rules for Breakout Sessions

12:00 noon   Lunch

1:00 p.m.   Parallel Breakout Sessions: First Four
               Challenges, Two Challenges per Group

2:30 p.m.   Break

2:45 p.m.   Two Groups on Same Challenges Meet and
               Reconcile Challenges and Lines of Inquiry

3:45 p.m.   Break

4:00 p.m.   Plenary Session to Present and Discuss Four
               Reconciled Challenges and Lines of Inquiry—
               Four 15-Minute Presentations; 30 Minutes for
               Discussion

5:30 p.m.   Wrap-up

Thursday, May 8, 2008

8:15 a.m.	Parallel Breakout Sessions: Challenges 5 through 8, Two Challenges per Group
9:45 a.m.	Break
10:00 a.m.	Two Groups on Same Challenges Meet and Reconcile Challenges and Lines of Inquiry
11:00 a.m.	Break
11:15 a.m.	Plenary Session to Present and Discuss Four Reconciled Challenges and Lines of Inquiry— Four 15-Minute Presentations; 30 Minutes for Discussion
12:45 p.m.	Lunch
1:30 p.m.	Plenary Session: Review of Challenges Regarding Urgency and Implementation
2:30 p.m.	Break
2:45 p.m.	Plenary Session: Workshop Outcomes and Report
4:00 p.m.	Adjourn

DRAFT CHALLENGES: MOVING TOWARD CRITICAL INFRASTRUCTURE SYSTEMS THAT ARE PHYSICALLY, ECONOMICALLY, SOCIALLY, AND ENVIRONMENTALLY SUSTAINABLE

- *Draft Challenge 1.* Ensure that the nation's critical infrastructure systems effectively support U.S. competitiveness in the global economy.
- *Draft Challenge 2.* Develop the critical infrastructure systems that support responsible U.S. energy independence.
- *Draft Challenge 3.* Upgrade, renew, replace, and provide new infrastructure systems to meet current and future requirements; improve reliability; improve performance and cost-effectiveness; promote equitably public safety, health, welfare, and social equity; and protect the environment.
- *Draft Challenge 4.* Optimize public- and private-sector investments in critical infrastructure systems and ensure adequate, long-term revenue streams for their operation, maintenance, and repair.
- *Draft Challenge 5.* Improve the reliability and resiliency of critical infrastructure systems to reduce the adverse impacts of human-made and natural disasters.
- *Draft Challenge 6.* Create a base of long-term support among users for infrastructure investments.
- *Draft Challenge 7.* Support innovation through the development and adoption of new approaches, technologies, and materials that have the potential to improve the delivery, quality, reliability, and sustainability of critical infrastructure services.
- *Draft Challenge 8.* Enhance international exchange and coordination of critical infrastructure systems approaches, services, components, and materials—with respect to finance, public and private ownership structures, regulations, and other factors.

D Summary of Workshop Outcomes

TOWARD SUSTAINABLE INFRASTRUCTURE SYSTEMS: WORKSHOP OUTCOMES

The discussions at the May workshop yielded many ideas and themes, or outcomes. The workshop outcomes are summarized below.

I. Introduction
   a. Many communities are facing challenges in maintaining and upgrading one or more of the basic services for their citizens (water, transportation, power, communications, wastewater).
      i. Reasons for the challenges
         1. Needs are changing.
         2. New approaches are available but not well known.
         3. Existing organizational structures impede coordination.
      ii. Result
         1. Each community is "reinventing the wheel."
         2. There is risk of potential suboptimal solutions across the full set of infrastructure services (e.g., corn as biofuel provides power but threatens water supply and land quality).

    iii. Opportunity
       1. Leverage current knowledge and experience across communities.
       2. Coordinate the development of new approaches.
       3. Coordinate implementation across communities, regions, and United States as a whole.
b. New approach: Infrastructure as—
    i. Service
       1. Provides critical functionality for civil society and commerce.
       2. Provides a basis for quality of life, well-being, and safety.
       3. Focuses on use rather than means of delivering.
    ii. Region
       1. Reflects actual system aspects of infrastructure (does not stop at community borders).
       2. Reflects links among communities for economic development, social equity, and environmental bearing capacity—at local, regional, national levels.
    iii. Interdependence
       1. Reflects functional and locational interdependence among infrastructure systems.
          a. For example:
             i. Water pumping and treatment requires power.
             ii. Power often requires water (for cooling, steam, etc.).
             iii. Power and telecommunications lines and water piping often run along transportation corridors.
       2. Reflects opportunities for further developments for sustainable infrastructure that explicitly take advantage of the integration of infrastructure systems to provide critical services.
          a. For example:
             i. Parking lots that generate electricity through photovoltaic coatings
             ii. Wastewater treatment plants that use biofuel cells to generate electricity
             iii. Localized gray water capture, treatment, and reuse with locally generated power

II. Conditions for Developing Sustainable Infrastructure Solutions
   a. Process
      i. Policy and planning
         1. Process for allocating funding
            a. Rational
            b. Transparent
            c. With respect to regional and national planning
         2. Champions for infrastructure services at community, regional, and national levels
         3. Flexible and adaptive policies with respect to economic, social, and environmental changes over time
         4. Balance of real cost to provide services and public good value (e.g., public health, commerce)
         5. Assessment of public "equity" in infrastructure assets—as reflected in property values, market activity, and so on
         6. Utilization of demand management (e.g., eliminate waste, increase efficiency)
         7. Partnership among private, public, and non-profit sectors
         8. Coordination/leverage of centralized and/or multi-nodal infrastructure systems with respect to
            a. Disaster resiliency and
            b. Flexibility in demand response
      ii. Decision making
         1. Transparency of infrastructure decision making
         2. Community, regional, and national reconciliation of infrastructure service needs and capacity—systemic, geospatial, strategic
         3. Decisions and solutions to enhance current infrastructure capacity with respect to community and regional economic development, environmental capacity, and social equity
         4. All-sector involvement (public, private, non-governmental organizations, community)

iii. Public dialogue and communication
1. Awareness of current infrastructure service capacity and vulnerabilities
2. Constant user feedback on condition, capacity, use, costs, and benefits of infrastructure systems
3. Recognized link of infrastructure services to economic development, social equity, and environmental regeneration
4. Local and regional dialogue on priorities, resources, and plans for infrastructure services (including resiliency and adaptiveness)
5. K-12 hands-on projects, simulation games, and other activities on the built and natural environments (i.e., infrastructure services and ecosystems)
6. Professional training with respect to current and emerging sustainable infrastructure services—local and regional capacity building
7. Executive sessions on the role of infrastructure services in organizational strategy and tactics
b. Structure
i. Financial
1. Alignment of cost and value of infrastructure services—structure of user, community, regional fees
2. Investment in new capacity to meet emerging and expected needs for infrastructure services
3. Investment in upgrades of existing infrastructure systems to meet current and expected needs
4. Clear designation of responsibilities, authorities, and financial means for delivery, operations, maintenance, and upgrade for infrastructure services over the lives of systems
5. All-sector involvement (public, private, nongovernmental organizations, community)
ii. Legal
1. Congruence in planning and operation with respect to the physical distribution of infrastructure systems

2. Regional integration of organizations that manage or oversee infrastructure systems and services
3. Insurance or warranty with respect to resiliency of infrastructure services for community and region
4. Mechanisms for international agreements, collaboration with respect to infrastructure services and ecosystem impacts

c. Performance
   i. Technological
      1. Investment in current, emerging, and "radical" technological developments
         a. Effectiveness
         b. Timescale
      2. Assessment and strategy for technology readiness
      3. International collaboration and information dissemination
      4. Modeling and real-time monitoring systems of infrastructure services (condition, capacity, use, cost, benefit, impacts)
   ii. Scientific Evidence and Metrics
      1. Performance criteria
         a. Physical
         b. Economic
            i. Cost
            ii. Benefit
            iii. Development
            iv. Secondary and tertiary impacts
         c. Social
         d. Environment and ecosystems
      2. Life-cycle analysis
         a. Timescale to reflect life of asset
         b. Multisector impacts
         c. Environmental footprint
         d. Secondary and tertiary impacts
      3. Analysis of systemic risks—especially with respect to interdependencies
         a. Reliability
         b. Robustness

4. Balance of human needs and bearing capacity of ecosystems
5. Trade-offs and priorities among and between infrastructure services and their underlying systems
6. All-hazards approach (cascading failures, differential vulnerabilities, etc.) with respect to potential disruptions, acceptable risks, climate change, and so on
7. Scale of infrastructure services and systems (e.g., spatial, organizational)

III. Existing Resources and Programs
   a. Previous U.S. programs
      i. New York Regional Plan
      ii. Interstate highway system
      iii. *Fragile Foundations* report
   b. Current national activities and programs
      i. Pending congressional bills
      ii. Regional agreements
   c. Local and state activities and programs
      i. Local (e.g., Cambridge Energy Alliance: nonprofit foundation, local government, private companies, universities and hospitals, citizens)
      ii. State (e.g., Hawaii renewable energy investment program)
   d. National laboratories
   e. Professional associations
      i. American Water Works Association and others
      ii. American Society of Civil Engineers' infrastructure report card

IV. Conclusions/Summary
   a. Focus on the future
   b. Focus on the possible
   c. Focus on starting the journey now, and learning as we go—
      i. Leverage current activities and programs and capabilities
      ii. Marshal knowledge, creativity, and engagement across all regions, sectors, and levels